T0185636

Lecture Notes in Physics

Volume 973

Founding Editors

Wolf Beiglböck, Heidelberg, Germany

Jürgen Ehlers, Potsdam, Germany

Klaus Hepp, Zürich, Switzerland

Hans-Arwed Weidenmüller, Heidelberg, Germany

Series Editors

Matthias Bartelmann, Heidelberg, Germany

Roberta Citro, Salerno, Italy

Peter Hänggi, Augsburg, Germany

Morten Hjorth-Jensen, Oslo, Norway

Maciej Lewenstein, Barcelona, Spain

Angel Rubio, Hamburg, Germany

Manfred Salmhofer, Heidelberg, Germany

Wolfgang Schleich, Ulm, Germany

Stefan Theisen, Potsdam, Germany

James D. Wells, Ann Arbor, MI, USA

Gary P. Zank, Huntsville, AL, USA

The Lecture Notes in Physics

The series Lecture Notes in Physics (LNP), founded in 1969, reports new developments in physics research and teaching-quickly and informally, but with a high quality and the explicit aim to summarize and communicate current knowledge in an accessible way. Books published in this series are conceived as bridging material between advanced graduate textbooks and the forefront of research and to serve three purposes:

- to be a compact and modern up-to-date source of reference on a well-defined topic;
- to serve as an accessible introduction to the field to postgraduate students and nonspecialist researchers from related areas;
- to be a source of advanced teaching material for specialized seminars, courses and schools.

Both monographs and multi-author volumes will be considered for publication. Edited volumes should however consist of a very limited number of contributions only. Proceedings will not be considered for LNP.
Volumes published in LNP are disseminated both in print and in electronic formats, the electronic archive being available at springerlink.com. The series content is indexed, abstracted and referenced by many abstracting and information services, bibliographic networks, subscription agencies, library networks, and consortia.
Proposals should be sent to a member of the Editorial Board, or directly to the responsible editor at Springer:

Dr Lisa Scalone
Springer Nature
Physics
Tiergartenstrasse 17
69121 Heidelberg, Germany
lisa.scalone@springernature.com

More information about this series at http://www.springer.com/series/5304

Martin Doubek • Branislav Jurčo •
Martin Markl • Ivo Sachs

Algebraic Structure of String Field Theory

Springer

Martin Doubek
Mathematical Institute
Faculty of Mathematics and Physics
(1982-2016) Dr. Doubek wrote this book
while at Charles University
Prague, Czech Republic

Branislav Jurčo
Mathematical Institute
Faculty of Mathematics and Physics
Charles University
Prague, Czech Republic

Martin Markl
Institute of Mathematics
Czech Academy of Sciences
Prague, Czech Republic

Ivo Sachs
Arnold Sommerfeld Center for Theoretical
Physics
Ludwig-Maximilian-University of Munich
München, Germany

ISSN 0075-8450 ISSN 1616-6361 (electronic)
Lecture Notes in Physics
ISBN 978-3-030-53054-9 ISBN 978-3-030-53056-3 (eBook)
https://doi.org/10.1007/978-3-030-53056-3

© Springer Nature Switzerland AG 2020
This work is subject to copyright. All are reserved by the Publisher, whether the whole or part of the material is concerned, specifically the rights of translation, reprinting, reuse of illustrations, recitation, broadcasting, reproduction on microfilms or in any other physical way, and transmission or information storage and retrieval, electronic adaptation, computer software, or by similar or dissimilar methodology now known or hereafter developed.
The use of general descriptive names, registered names, trademarks, service marks, etc. in this publication does not imply, even in the absence of a specific statement, that such names are exempt from the relevant protective laws and regulations and therefore free for general use.
The publisher, the authors,. and the editors are safe to assume that the advice and information in this book are believed to be true and accurate at the date of publication. Neither the publisher nor the authors or the editors give a warranty, expressed or implied, with respect to the material contained herein or for any errors or omissions that may have been made. The publisher remains neutral with regard to jurisdictional claims in published maps and institutional affiliations.

This Springer imprint is published by the registered company Springer Nature Switzerland AG.
The registered company address is: Gewerbestrasse 11, 6330 Cham, Switzerland

Dedicated to the memory of Martin Doubek who died tragically on 29th of August 2016.

Preface

The subject of these lecture notes is modular operads and their rôle in string field theory. We believe that the approach to string field theory based on homotopy algebras and their operadic origin can be of interest to both theoretical physicists and mathematicians. A mathematician can perhaps, starting from operads and homotopy algebras related to them, find some conceptual explanation for the algebraic structure of string field theory. A mathematically oriented physicist, starting from string field theory, may find useful and appealing mathematical language and tools of the theory of operads. We hope that this text will provide some inspiration to both of them and thus enhance mutual interaction and progress in both areas.

Our presentation of string field theory is neither a textbook nor it provides a proper introduction to the subject. This was not our aim. As already mentioned above, we tried to present it, without going to details, from the perspective of homotopy algebras and their operadic origin. Concerning operads, the presentation can be used as an introduction to modular operads, their Feynman transform and corresponding homotopy algebras.

In Part I, we start with a description of string field theory that should be familiar to physicists and develop it in a way which makes the appearance of homotopy algebras transparent. In this approach, Zwiebach's construction of a string field theory naturally emerges as a composition of two morphisms of particular odd modular operads. In more traditional terms, these morphisms correspond to a decomposition of a moduli space of Riemann surfaces and to conformal field theory, respectively. A mathematically rigorous description of operads is presented in detail in Part II. Whenever possible we comment on the connection between the two parts by at the end of the respective sections.

A more detailed description of the structure of the book and the content of the individual chapters is given in the introduction. In Part I, written by B. Jurčo and I. Sachs, we comment at the end of each chapter on literature, which either we followed in our exposition or where an interested reader can find further details. In Part II, written by M. Doubek and M. Markl, the references are listed at the end of each chapter.

I.S. and B.J. would like to thank Maxim Grigoriev for helpful comments on Part I of the book and Kai Cieliebak, Ted Erler, Korbinian Muenster, and Sebastian Konopka for valuable discussions that helped to develop the ideas that are presented in this book. I.S. would like to acknowledge the hospitality of Clare

Hall, Cambridge where most of Part I was written and the DFG Excellence cluster ORIGINS. Research of B.J. was supported by GAČR Grant EXPRO 19-28268X. M.M. acknowledges the support of Praemium Academiae, grant GAČR 18-07776S and RVO: 67985840 during the last stages of his work on the book.

Prague, Czech Republic — Branislav Jurčo
Prague, Czech Republic — Martin Markl
Munich, Germany — Ivo Sachs

Contents

About the Authors

Martin Doubek graduated from the Faculty of Mathematics and Physics of the Charles University in Prague. He wrote his Ph.D. thesis under the supervision of Martin Markl at the Mathematical Institute of the Czech Academy of Sciences and defended it in 2011. The promising career of this talented young mathematician was terminated in 2016 by his tragic death in a traffic accident.

Branislav Jurčo graduated from the Faculty of Nuclear Sciences and Physical Engineering of the Czech Technical University. He defended his Ph.D. thesis, supervised by Jiří Tolar, in 1991. He was a Humboldt Fellow at TU in Clausthal and MPIM in Bonn. His postdoctoral experience also includes stays at CERN, CRM in Montreal, and LMU in Munich. He is currently an Associate Professor at the Faculty of Mathematics and Physics of the Charles University in Prague. His research focuses on applications of higher algebraic and geometric structures in theoretical and mathematical physics.

Martin Markl graduated from the Faculty of Mathematics and Physics of the Charles University in 1983 and defended his Ph.D. thesis, written under the supervision of Vojtěch Bartík, in 1987. He was influenced at the early stage of his research career by Jim Stasheff during his repeated visits at the University of North Carolina at Chapel Hill. He is currently a senior research fellow of the Mathematical Institute of the Czech Academy of Sciences in Prague. His research is focused on homological algebra, geometry, and applications to mathematical physics.

Ivo Sachs graduated in 1991 from the faculty of physics of the ETH in Zurich and defended his Ph.D. in 1994 under the supervision of Andreas Wipf. In 2001, he became a lecturer at the School of Mathematics at Trinity College, Dublin and, later on, was awarded a professorship in theoretical physics at the Ludwig-Maximilians-University in Munich in 2003. His main achievements are in quantum field theory, the structure of black holes, and string field theory.

1 Introduction

Loosely speaking, a modular operad can be pictured as a set of vertices, or corollas, labeled by the number of legs n and a further integer g interpreted as the genus. Two corollas can be composed. For each pair of legs (one from the first corolla, another from the second one), the composition is done by joining the legs and contracting the resulting edge. This composition decreases the total number of legs by two and is additive with respect to the genera. Similarly, for each pair of legs of the same corolla, there is a self-composition consisting of joining the legs and contracting the resulting edge. Such a self-composition reduces the number of legs by two and increases the genus by one. Further, corollas admit an action of the symmetric group, by permuting the legs of a corolla. To define the structure of a modular operad (in the category of differential graded vector spaces) we need a prescription associating to each corolla a dg (differential graded) vector space. Also, we need a prescription transferring the compositions and the action of the symmetric group from corollas to the associated vector spaces in a natural way. These prescriptions determine the particular kind of a modular operad we are dealing with, e.g., modular commutative operad, modular associative and so forth.

The Feynman transform associates to a modular operad an odd modular operad in a way reminiscent of constructing Feynman graphs in quantum field theory. Here, "odd" refers to the operadic compositions, which have now degree one. Roughly speaking, Feynman transform is a free operad generated by a properly suspended dual and a differential formed using the duals of the operadic compositions of the original modular operad. Since as an odd operad the Feynman transform is generated by corollas, it can be described in terms of graphs with external legs and any numbers of loops resulting from concatenation of corollas, similarly as Feynman graphs are composed from the interaction vertices.

An important example of an odd modular operad is the endomorphism operad associated with a differential graded vector space equipped with an odd symplectic form. An algebra over an odd modular operad is a morphism from it to the endomorphism operad, i.e. its representation. The algebra resulting in this way

© Springer Nature Switzerland AG 2020

M. Doubek et al., *Algebraic Structure of String Field Theory*, Lecture Notes in Physics 973, https://doi.org/10.1007/978-3-030-53056-3_1

from the Feynman transform of the modular commutative operad is the loop homotopy algebra (aka quantum L_∞-algebra). Such an algebra can equivalently be characterized by a solution to the quantum Batalin–Vilkovisky (BV) master equation, i.e., by Maurer–Cartan elements of the corresponding BV algebra. The origin of the BV operator and the BV bracket can be traced back to the operadic self-compositions and compositions, respectively. Obviously, this is one of the reasons why operads and the corresponding algebras are relevant to physics. More generally, a morphism from the Feynman transform of any modular operad to a general odd modular operad can be described by a solution to a proper generalization of the quantum BV equation.

All of these constructions contain a genus zero part, due to the forgetful functor from the category of modular operads to the category of cyclic operads. One simply forgets about loops. Hence, the dg-vector spaces associated with corollas with non-zero genus are trivial as well as the operations corresponding to self-contractions of corollas. This functor has an adjoint, the modular completion (or envelope), which roughly speaking freely adds loops. The resulting (cyclic) homotopy algebras (e.g., L_∞- or A_∞-algebras) can then be described by solutions to the corresponding classical master equations.

All this is nicely illustrated by string field theory. For simplicity, let us consider closed strings. One starts with the modular commutative operad (the modular envelope of the cyclic commutative operad) and its Feynman transform. The theory of closed strings provides us with two natural odd modular operads. One is formed by the singular chain complex on the moduli space of closed Riemann surfaces with punctures and with the operadic compositions induced from (twisted) sewing and self-sewing of surfaces. The other one is the endomorphism operad over the state space of the (first quantized) closed string with the differential given by the BRST (Becchi-Rouet-Stora-Tyutin) operator and the odd symplectic form induced by the BPZ (Belavin-Polyakov-Zamolodchikov) pairing.

A morphism from the Feynman transform of the modular commutative operad to the first one gives a BV structure on the moduli space of Riemann surfaces and a solution to the corresponding quantum BV equation. This solution describes the decomposition of the moduli space and hence tells us what the geometric vertices are. This is the geometric background independent part of the construction of closed string field theory in Zwiebach's approach. It is then composed with a morphism of odd modular operads, provided by conformal field theory, going from the moduli space operad to the endomorphism operad associating algebraic vertices to the geometric ones. This is the background dependent part of the construction. This composition describes a particular morphism from the Feynman transform of the modular commutative operad to the endomorphism operad, i.e., a solution to the quantum BV equation, describing the algebraic vertices forming the quantum BV action of the closed string field theory. Obviously, the vertices are labeled by their valence n and the genus g. The algebraic vertices are degree zero graded symmetric functions on the string state space. Using the odd symplectic form, one can equivalently use the string operations, i.e., degree one graded symmetric multilinear maps from the state space to itself. Hence, we have a description in terms of a collection of $n-1$-ary brackets of degree one labeled by genus g. The genus zero part of the construction describes the classical closed string field theory.

This picture remains valid for the full quantum open-closed string field theory. One has to use a two-colored modular operad combining the modular commutative and associative operads, consider the moduli space of bordered Riemann surfaces with punctures in the interior as well as on the boundaries, and the open-closed conformal field theory.

When described in terms of higher brackets, a loop homotopy algebra can be viewed as a particular example of an IBL_∞ (homotopy involutive Lie bi-) algebra. For this, we view the BV operator Δ as an operation with zero input and two outputs, i.e., as a particular cobracket. The algebraic properties of the brackets and the cobracket are (after a dualization) encoded in the square-zero condition on the full quantum BRST operator $\Delta + \{S, -\}$. Here, S is the quantum BV action and $\{-, -\}$ the BV bracket. IBL_∞-algebras are algebras over the cobar complex of the terminal properad. Roughly speaking, we consider corollas as before but now their legs are oriented, they can be either outputs or inputs. The operations can now be pictured by simultaneously joining several outputs of one corolla with inputs of an another one followed by a contraction of new edges created in the process. Since the simultaneous joining of legs of two corollas increases the genus, we do not need an analogue of the self-composition from modular operads any more. In this book we describe IBL_∞-algebras, but refrain from introducing properads. The related constructions are analogous to the case of operads but more complex. Nevertheless, the approach using the language of IBL_∞-algebras is fruitful and can be used for an algebraic description of the open-closed string theory, leading to its formulation analogous to Kontsevich's formality.

We start with an introduction of the concepts relevant to the BV action as it arises in string field theory in Part I, while in Part II we focus on the theory of (cyclic, modular, odd modular) operads and morphisms between these which, in turn, correspond to the homotopy algebras introduced in the physics context of Part I. Also, a concise mathematical description of IBL_∞-algebras is given in Part II. A more informal parallel description of these in a form directly used in our description of the open-closed string field theory is provided in the appendix to Part I. Whenever possible we will comment on the connection between these two parts at the end of the respective sections.

Structure of the Book In the first chapter, we formulate the relativistic scalar field within the BV formalism starting from the world-line formulation for a relativistic particle. This serves primarily as a toy model and a motivation for the analogous construction in string field theory. In this approach, the appearance of differential graded algebras and homotopy algebras within the BV formalism becomes very natural. As explained in detail in Part II, the homotopy algebras of quantum field theory and string field theory are representations of operads. For example, as already mentioned above, loop homotopy algebras which are directly related to the usual BV formalism are described as operad morphisms from the Feynman transform of the modular envelope of the cyclic commutative operad to the endomorphism operad.

Operads related to moduli spaces will not appear in this chapter. Though it seems natural to consider moduli spaces related to metric graphs, this would take us far beyond the scope of the present lecture notes.

In Chap. 2, we generalize the above constructions to closed strings. This naturally leads to the construction of BV algebras and their Maurer–Cartan elements through the decomposition of the relevant moduli spaces of Riemann surfaces. Combined with the morphism from the moduli BV algebra to the closed string BV algebra provided by the conformal field theory, this decomposition gives the vertices of the corresponding string field theory action. The whole construction is mathematically interpreted again in terms of a representation of the Feynman transform of the modular commutative operad described in Part II. We also include a section on the uniqueness and background independence of the resulting algebraic structure, which is an important issue in string theory but can be bypassed by readers who are primarily interested in the operadic/algebraic aspects of string field theory only.

In Chap. 3, we introduce open strings. Here, in analogy with the point particle, the algebraic structure simplifies to that of a differential graded algebra (DGA). Nonequivalent deformations of classical open string theory are governed by its cyclic cohomology. Non-trivial elements are realized as A_∞-algebras that come with it. According to general theory described in Part II, A_∞-algebras and their cyclic versions are interpreted as representations of the cobar complex of the Ass-operad and its cyclic version, respectively.

Part I then closes with Chap. 4, which gives an account of the open-closed homotopy algebra. The chosen algebraic description is that using involutive bi L_∞- (IBL_∞-) algebras which are revisited in the appendix to this chapter and again, from an alternative angle and more rigorously, in the last section of Part II. As already mentioned, this approach would need to introduce properads, which is far beyond the scope of this introductory lecture notes. The open-closed homotopy algebra has also an operadic description in terms of a colored operad (open-closed operad combining the modular commutative and associative operads) which will, however, also not be covered in this book. This is, however, a straightforward extension.

The main idea behind the presentation and the structure of this book is perhaps best summarized by the commutative diagram of odd modular operads in Fig. 1.1, leading to a construction of a quantum string field theory. Let us specify it in the case of the closed bosonic quantum string field theory, the example discussed most thoroughly in these lecture notes. Starting from the Feynman transform of the modular commutative operad, the horizontal arrow determines a Maurer–Cartan element in the BV algebra of chains on the moduli space of closed Riemann surfaces with punctures, i.e., a decomposition of the moduli spaces determining the geometric vertices. The vertical arrow stands for a morphism from the BV algebra of chains on this moduli space to the endomorphism operad determined by the closed conformal field theory. In particular, it maps a Maurer–Cartan element (the geometric vertices) to the Maurer–Cartan element (the algebraic vertices) that defines a gauge invariant bosonic closed quantum string field theory action. Hence, the diagonal arrow determines a particular solution to the algebraic Batalin–Vilkovisky master equation on the BV algebra of functions on the underlying graded

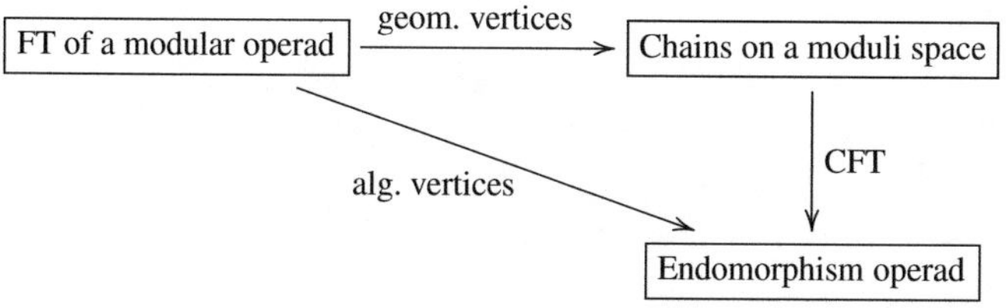

Fig. 1.1 Construction of string field theory in terms of morphisms of (odd modular) operads

differential vector space V, the state space of the closed conformal field theory, i.e., of the fist quantized closed bosonic string.

As already mentioned above, the bosonic quantum open-closed string theory can be constructed similarly. Furthermore, we believe that also superstring field theories could in principle be constructed in this way, here the difficult part is the horizontal arrow. It is not obvious how to mimic Zwiebach's construction of geometric vertices to the case of super Riemann surfaces. Finally, since any quantum field theory can be formulated as a BV theory, it should allow for an interpretation/construction according to our commutative triangle. Probably, the moduli spaces of tropical curves and world line formulation of quantum field theory (QTF) should lead to this as well.

Part I
Physics Preliminaries

The standard formulation of perturbative string theory consists of a set of rules to compute scattering amplitudes for a collection of n particles—excitations of a string—typically propagating on a D-dimensional Minkowski space-time M. In the simplest case of tree level scattering amplitudes, this prescription involves an integration over the moduli space of spheres with n punctures (or disks with n punctures on its boundary for open strings) and provides a very efficient procedure to compute scattering amplitudes. On the other hand, comparing this with the approach taken for point particles the situation in string theory seems somewhat incomplete. Indeed, for point particles one starts with an action principle for a set of fields and then obtains the tree level scattering amplitudes by solving the equations of motions deriving from this action. Since the various string excitations ought to be interpreted as particles, one would hope to be able to apply the same formalism for the scattering of strings. The aim of string field theory is precisely to provide such an action principle so that the set of rules to compute scattering amplitudes for strings derive from this action. Since the string consists of an infinite linear superposition of point particle excitations, one would expect that such an action may be rather complicated. Yet the first construction of a consistent classical action of interacting open strings has a remarkably simple algebraic structure of a differential graded associative algebra $(V, *, d)$ together with an odd symplectic form compatible with d and $*$. Such a structure is called a cyclic differential graded associative algebra.

It turns out that this simple theory does not lead to a consistent quantum theory without the inclusion of closed strings. However, closed strings require a more elaborate construction, which nevertheless has a similarly appealing algebraic structure. For this, one necessarily needs all closed string loops. Therefore, one starts with a decomposition of the relevant moduli space of closed punctured Riemann surfaces Σ together with coordinate curves around each puncture into elementary vertices and propagators. The condition that this moduli space is covered exactly once by the Feynman diagrams built from this data implies that the geometric vertices satisfy a Batalin–Vilkovisky (BV) master equation. This equation encodes information about the decomposition of the relevant moduli space. In a next step one endows the punctures of Σ with an extra structure by attaching a vector space

V to each puncture. This vector space could be the deRham complex of the target manifold, as is the case for the topological string, or the quantum mechanical Hilbert spaces of physical degrees of freedom such as photons or gravitons. The geometric BV master equation is then transferred to V by means of a two-dimensional conformal field theory defined on the geometric vertices. This conformal field theory furthermore endows V with an inner product, more precisely, with a symplectic form. This provides the data needed to define a BV action on V, that is a field theory action of closed strings.

It turns out that, at genus zero, the vertices of this BV action satisfy the axioms of a cyclic L_∞-algebra, which is a differential graded Lie algebra up to homotopy. Thus, we find that the mathematical structure of classical closed string field theory is simply that of a cyclic L_∞-algebra supported by the vector space V of closed string states together with a symplectic form. Similarly, at genus zero, any consistent covariant open string field theory carries a structure of a cyclic A_∞-algebra, that is, a differential graded associative algebra up to homotopy.

We already mentioned above that open string field theory is not consistent at the quantum level without inclusion of closed strings. This raises the question about the algebraic structure of the combined system of open and closed string field theory. At genus zero, it is that of an open-closed homotopy algebra which is a realization of an L_∞-morphism that maps closed string states to open string vertices. This map turns out to be a quasi-isomorphism, that is, there is an isomorphism between physical closed string states and classically consistent open string field theories.

If we include closed string loops, the elementary geometric vertices are realized by genus g Riemann surfaces with punctures. The quantum master equation then involves the BV operator Δ. This originates from the corresponding operation acting on a geometric vertex, which increases the genus of Σ by one by glueing two punctures with a zero length propagator and thus comes with a power of $\hbar$ in the BV equation. If the algebraic vertices are interpreted as functions on the (graded symmetric) tensor products of the physical Hilbert space (denoted by V above) then Δ can be implemented as the inverse of the symplectic form ω on V. The algebra realized by these functions subject to the BV master equation is the loop homotopy algebra (or quantum L_∞-algebra). Since closed string field theory is perturbatively consistent, this is the algebraic structure of the complete quantum field theory of closed strings.

The quantum open-closed homotopy algebra has an analogous physical interpretation as the classical (genus zero) one. For this the natural framework is that of IBL_∞-algebras. The quantum open-closed vertices realize an IBL_∞-morphism that maps closed string states to open string vertices of higher genus. We will discuss all this in the present lecture notes from both mathematics and physics points of view.

Relativistic Point Particle

2

As briefly explained in the introduction, the BV formalism plays the central role in the construction of a gauge invariant action for string field theory. By now, there are many excellent texts on the BV formalism. We will thus not present a detailed account. However, the construction of a BV action in string field theory is rather different from that in gauge theories. In some ways it goes backwards! Indeed, in the original BV construction, the starting point is a gauge invariant action (or boundary condition). The gauge theory is then extended by addition of further fields (ghost, ghosts for ghosts, etc.) and anti-fields. The action gets extra terms which contain new fields and anti-fields. Expanding the so obtained extended action around a critical point naturally induces an L_∞-algebra structure. In contrast, in string field theory, the initial action or boundary condition is generically unknown. Nevertheless, starting from the world-sheet description which is a convenient formulation of string theory, and decomposing the scattering amplitudes into irreducible components, one directly constructs the perturbative expansion of the BV action around a critical point with all anti-fields included.

It turns out that a simplified version of the algebraic structure described in the introduction is already present for point particles. We will therefore begin with the construction of the BV action for a spinless particle starting from the world-line formalism. Of course, for that system a consistent field theory can easily be constructed by other means as it is done in the standard textbooks of quantum field theory. However, for illustration of the formalism at work in string theory, including the algebraic structure and its operadic description, this approach seems appropriate.

2.1 BV Action for the Point Particle

We consider a single point particle propagating on a non-compact manifold M with a pseudo-Riemannian metric g. The world-line of the particle defines a parametrized curve $\phi : [\tau_i, \tau_f] \to M$. The basic object in our construction of a space-time action

© Springer Nature Switzerland AG 2020

M. Doubek et al., *Algebraic Structure of String Field Theory*, Lecture Notes in Physics 973, https://doi.org/10.1007/978-3-030-53056-3_2

for the point particle is the world-line action functional

$$I[\phi, e] = \frac{1}{2}\int_{[\tau_i,\tau_f]} \left(\frac{1}{e}g(\dot{\phi}, \dot{\phi}) - m^2 e\right) d\tau\,, \tag{2.1}$$

where $e(\tau)$ is a non-dynamical einbein on the world-line and the vector field $\dot{\phi} = \frac{d}{d\tau}\phi \in TM$ is tangent to the curve ϕ. As a consequence of the reparametrization invariance, $\phi(\tau) \to \phi(f(\tau))$, this action is invariant under

$$\delta\phi = \frac{\dot{\phi}}{e}\epsilon(\tau)\,, \qquad \delta e = \dot{\epsilon}(\tau)\,, \tag{2.2}$$

for $f(\tau) = \tau - \frac{\epsilon(\tau)}{e(\tau)} + O(\epsilon^2)$, where ϵ is a smooth function with $\epsilon(\tau_f){=}\epsilon(\tau_i){=}0$. In particular, with a suitable choice of $f(\tau)$ we can set $e(\tau)$ to be a constant. On the other hand, elimination of e by solving the corresponding constraint equation $\frac{1}{e^2}g(\dot{\phi}, \dot{\phi}) + m^2 = 0$ reproduces the usual geodesic length

$$I[\phi] = m\int_{\tau_i}^{\tau_f} \sqrt{-g(\dot{\phi}, \dot{\phi})}\, d\tau\,,$$

which is the familiar action for a relativistic point particle.

For quantization, the action (2.1) is, however, more suitable since it is polynomial in the world-line field ϕ. We first consider path integral quantization which is formally defined by replacing classical trajectories by a functional integral over paths with evolution kernel

$$K(\tau_f, \tau_i, \phi_f, \phi_i) \overset{form}{=} \int D[\phi, e]\, e^{\frac{i}{\hbar}I[\phi,e]}\,. \tag{2.3}$$

Here we assume fixed boundary conditions $\phi(\tau_i) = \phi_i$, $\phi(\tau_f) = \phi_f$. Due to the invariance (2.2) the path integral measure in (2.3) is degenerate. This can be remedied with the help of the Faddeev–Popov BRST procedure where one starts with an action functional on a super manifold M

$$I[\phi, e, b, c, \bar{\pi}] = \int_{\tau_i}^{\tau_f} \left(\frac{1}{2e}g(\dot{\phi}, \dot{\phi}) - \frac{m^2 e}{2} + ib\dot{c} + \bar{\pi}(e - \hat{e})\right) d\tau\,, \tag{2.4}$$

where $\hat{e}$ is a reference einbein which can be set to be one, $\hat{e} = 1$, by rescaling τ. The maps

$$\{\phi, \bar{\pi}, b, c\} : [\tau_i, \tau_f] \to M \tag{2.5}$$

from the world-line to a super manifold form a graded commutative algebra under point-wise multiplication. In physics, the $\mathbb{Z}$ grading is often referred to as the ghost number gh, with $gh(\phi, e, \bar{\pi}) = 0$ and $gh(c) = -gh(b) = 1$, respectively. In this chapter, we will identify this ghost number with the degree. The Lagrange multiplier field $\bar{\pi}$ localizes the functional integral over e which is sufficient to remove the degeneracy of the path integral measure. The role of the (b, c) ghost system is to cancel the Jacobian arising from the reparametrization (2.2).

The constraint action (2.4) has a nilpotent (BRST) symmetry

$$\delta_{BRST} e = -i\dot{c}\,, \quad \delta_{BRST}\phi = -i\frac{\dot{\phi}c}{e}\,, \quad \delta_{BRST} b = \bar{\pi}\,, \quad \delta_{BRST} c = \delta\bar{\pi} = 0\,,$$

as can be most easily seen by noticing that

$$I[\phi, e, b, c, \bar{\pi}] = I[\phi, e] - \int \delta_{BRST} b(e - \hat{e}).$$

Clearly, $\delta^2_{BRST} = 0$, so that δ_{BRST} defines a cohomological vector field on the graded space of trajectories (2.5). The addition of the BRST exact term in $I[\phi, e, b, c, \bar{\pi}]$ localizes the functional integral at the critical points of the BRST exact piece. Indeed, upon elimination of $\bar{\pi}$ the constraint $e \equiv 1$ is enforced so that the kernel

$$K(\tau_f, \tau_i, \phi_f, \phi_i) \equiv \int D[\phi, b, c]\; J\, e^{\frac{i}{\hbar}\int_{\tau_i}^{\tau_f}\left(\frac{1}{2}g(\dot{\phi},\dot{\phi}) - \frac{1}{2}m^2 + ib\dot{c}\right)d\tau} \tag{2.6}$$

is well defined. Here,

$$J = \frac{1}{\det'(\frac{i}{\hbar}\partial_\tau)}$$

is included for later convenience. The prime $'$ is to highlight the fact that the constant mode corresponding to the translation in proper time has not been included in the above path integral measure. It will be included below in Eq. (2.11). The reduced action (2.6) still enjoys a residual symmetry

$$\delta_{BRST}\phi = -i\dot{\phi}c\,, \quad \delta_{BRST} b = \frac{1}{2}\left(g(\dot{\phi}, \dot{\phi}) + m^2\right). \tag{2.7}$$

In the canonical (as opposed to path integral) quantization of the reduced action (2.6), $\{p, \phi, c, b\}$ form a graded operator algebra with non-vanishing, graded, equal proper-time commutation rules

$$[\phi, p] = i\mathbb{1} \qquad \text{and} \qquad [b, c] = \mathbb{1}\,, \tag{2.8}$$

acting on a graded vector space V over $\mathbb{C}$ spanned by the quantum-mechanical states ψ of our spinless particle. Here and in what follows we set $\hbar = 1$. To construct an irreducible module for the above canonical commutation relations, we start with representing the operator ϕ on functions as the multiplication by the cartesian coordinates ϕ^i and p as the derivative

$$p = \frac{1}{i}\nabla_\phi \,.$$

The degree zero part of the module V is then naturally identified with the vector space of test functions on M with suitable regularity and integrability properties. To continue, we realize the anticommuting (b, c)-system on V in analogy to the (p, ϕ)-system by taking c to act through multiplication by c and

$$b = \frac{\partial}{\partial c}\,. \tag{2.9}$$

This completes the construction of V as a module over the Grassmann numbers $\mathbb{C}_{\mathbb{Z}_2} = \mathbb{C}_0 \oplus \mathbb{C}_1$, where $\mathbb{C}_0$ resp. $\mathbb{C}_1$ represent the commuting, respectively, anticommuting numbers. Altogether, V is the tensor product of the space of wave functions of a scalar particle with the two-dimensional Fock space of the (b, c)-system. A generic vector $\psi \in V$ is of the form $f_0(\phi) + f_1(\phi)c = \Psi^i e_i$, where $\{e_i\}$ is a homogeneous basis of the vector space V. f_0 and f_1 are $\mathbb{C}$- and Grassmann-valued functions, respectively, interpreted as space-time fields. The degree of Ψ^i is such that Ψ has total degree 0.

The BRST symmetry (2.7) is then generated through commutation with

$$Q = c\,\mathcal{H}, \quad \mathcal{H} = \frac{1}{2}\left(g^{-1}(p,p) + m^2\right). \tag{2.10}$$

Remark 2.1 In the BV formalism which we will review below, fields of degree zero are interpreted as classical fields, while fields of degree -1 are referred to as antifields, see also Sect. 2.3. Also, sometimes it will be convenient to write $\delta(c)$ instead of c. This is possible due to c being of odd degree.

Remark 2.2 To derive the correct BRST transformations in this way, one has to use the equation of motion $g(\dot{\phi}, \cdot) = p$. Alternatively, one can work in the phase space where

$$I[\phi, b, c, \bar{\pi}] = \int_{\tau_i}^{\tau_f} \left(p\dot{\phi} - \frac{1}{2}(g^{-1}(p,p) + m^2) + ib\dot{c}\right) d\tau\,,$$

which, in turn, is invariant under

$$\delta_{BRST}\phi = -ip(c)\,, \qquad \delta_{BRST}b = \frac{1}{2}\left(p^2 + m^2\right), \quad \text{and} \quad \delta_{BRST}c = \delta_{BRST}p = 0\,,$$

now generated by (2.10) without using equations of motion.

The action of Q on V is unambiguously determined by the representation (2.8) and (2.9). Furthermore, $Q\psi = 0$ enforces the *physical* or *on-shell* condition. More precisely, the physical states are usually identified with $H|_{\deg 0}$, where $H = \mathrm{coh}(Q, V)$, but in this book we will sometimes ignore the restriction to deg 0 and refer to physical states simply as to elements of H and use sometimes the notation V_{phys} or V_P for it.

Given our construction of V above, $\psi \in V$ is physical or *on shell* if $\mathscr{H}\psi_0 = 0$. Note that, since $\mathscr{H} = [Q, b]$, the world-line Hamiltonian $\mathscr{H}$ for this dynamical system is Q-exact.

Returning to the path integral, for a given interval $\Delta\tau = \tau_f - \tau_i$, the kernel $K(\Delta\tau, \phi_1, \phi_0)$ has the physical interpretation of the quantum-mechanical propagation in proper time between initial and final state of the particle by integrating the kernel against the wave functions $\psi(\phi, c)$ as

$$K(\Delta\tau, \psi_2, \psi_1) = \int dc\, d\phi_2\, d\phi_1\; \bar{\psi}_2(\phi_2, c)\psi_1(\phi_1, c)\; K(\Delta\tau, \phi_2, \phi_1)\,. \tag{2.11}$$

Here, the integral $\int dc$ over the constant c-ghost arises from fixing the invariance of the world-line path integral measure under proper-time translations. For $\Delta\tau \to 0$, the functional integral kernel (2.6) reduces to a delta function[1]

$$\delta(\phi_2 - \phi_1) \tag{2.12}$$

thus (2.11) defines a degree -1 symplectic form

$$\langle\psi_2, \psi_1\rangle \equiv K(0, \psi_2, \psi_1) = \int dc d\phi\; \bar{\psi}_2(\phi, c)\psi_1(\phi, c)\,. \tag{2.13}$$

Note that, due to the presence of the c-zero mode, the paring is between subspaces of V with different degree. This may seem counter intuitive but it is, in fact, a common feature in BV quantization.

We now have all ingredients needed to define an action functional on V,

$$S[\psi] = \frac{1}{2}\langle\psi, Q\,\psi\rangle\,. \tag{2.14}$$

This action reproduces the correct on-shell, or physical state condition from the variational principle for ψ when evaluated at degree 0.

Interactions between scalar particles can be included by considering the propagation of the first particle in the potential of the second as in Fig. 2.1. If we denote by $\psi_2(\phi)$ the wave function of the second particle, then the world-line of the first particle couples to the latter through $\psi_2(\phi(s))$ where s is the proper time of the first

[1]Strictly speaking this holds only for Riemannian signature on M and should be suitably defined with an $i\epsilon$ prescription as known from quantum field theory textbooks.

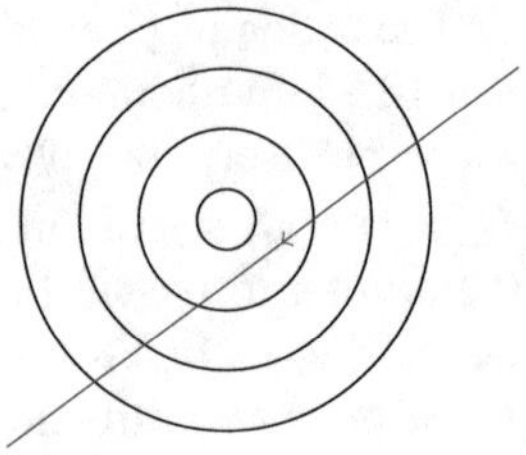

Fig. 2.1 Sketch of the world-line of the first particle (straight line) in the background of the second represented by circles in analogy with to propagation through a water wave

particle. Therefore, the correct generalization of the evolution kernel (2.11), linear in ψ_2, is obtained by replacing the right-hand side of (2.11) by

$$\int_{\Delta\tau} ds \int dc\, d\phi_f\, d\phi_i\; \bar{\psi}_3(\phi_f) \int D[\phi, b, c]\; e^{iI[\phi,b,c]}\; \delta(c(s))\, \psi_2(\phi(s))\, \psi_1(\phi_i) \tag{2.15}$$

$$= \int_{\Delta\tau} ds \int dc\, d\phi_f\, d\phi_i d\phi\; \bar{\psi}_3(\phi_f)\; K(\tau_f, s, \phi_f, \phi)\, c(s)\, \psi_2(\phi) \times K(s, \tau_i, \phi, \phi_i)\; \psi_1(\phi_i)\,,$$

where we used again that $\delta(c(s)) = c(s)$ in the second line and omitted the ghost c from the arguments. The insertion of $\delta(c(s))$ in the above expression is required to ensure that the only reparametrizations that are gauge fixed are those which preserve the interaction point. If we then strip off the free propagation which amount to letting $\Delta\tau \to 0$ in the evolution kernel, this defines a product on V,

$$* : V \otimes V \to V\,, \quad \psi \otimes \psi \mapsto c\psi^2\,,$$

which is commutative, associative, and of degree one. It is not hard to see that (2.15) can be reproduced upon adding a cubic term to (2.14), i.e.,

$$S[\psi] = \frac{1}{2}\langle\psi, Q\,\psi\rangle + \frac{1}{3}\langle\psi, \psi * \psi\rangle\,. \tag{2.16}$$

The algebraic structure of the action (2.16) is that of a nilpotent abelian differential graded (dg) algebra $(V, Q, *)$, together with a non-degenerate symplectic form ω. The relations are

$$Q^2 = *^2 = Q* = *Q = 0\,. \tag{2.17}$$

Finally, let us comment on the BV action corresponding to (2.16). Formally, this action is invariant under $\psi \to \psi + Q\lambda$ where λ has degree -1. The idea of the

BV formalism is to generate this transformation classically, through a canonical transformation

$$\delta\psi = \delta\Psi^i e_i = \{S, \Psi^i\} e_i\,, \tag{2.18}$$

where the BV bracket $\{-,-\}$ is defined through

$$\{A, B\} = \frac{\partial_l A}{\partial \Psi^i}\omega^{ij}\frac{\partial_r B}{\partial \Psi^j}\,. \tag{2.19}$$

Here $\partial_{l/r}$ denote the left/right derivatives and ω^{ij} is the inverse of the *odd* symplectic form $\omega_{ij} d\Psi^i \wedge d\Psi^j$, where Ψ^i is an odd or even coordinate. Furthermore,

$$\omega_{ij} = \omega(e_i, e_j) \equiv (-1)^{deg(e_i)}\langle e_i, e_j\rangle\,. \tag{2.20}$$

The BV action corresponding to (2.16) is then given by

$$S[\psi] = \frac{1}{2}\omega(\psi, Q\,\psi) + \frac{1}{3}\omega(\psi, \psi * \psi) \tag{2.21}$$

but without any restriction on the degree of ψ. This is not a generic feature of BV actions. However, it will be the case in the examples relevant for this book. The invariance of the action is then encoded in the classical BV equation

$$\{S, S\} = 0\,. \tag{2.22}$$

The familiar cubic action for a scalar field is obtained by restricting ψ to the degree zero subspace of V.

Remark 2.3 The alert reader noticed that the gauge transformations are in fact trivial in the present case since there are no degree -1 fields in V for the point particle. Thus, the ghost system is redundant for the action (2.18) on V and there is a much simpler description of the spinless relativistic particle by dropping the c ghost altogether in the definition of the action (2.18) and working directly with the inner product given on the right-hand side of (2.13) without the ghost contribution. Nevertheless we will continue to use this "redundant" notation because it serves as a simple and useful illustration for the string.

Remark 2.4 Throughout this chapter we have considered propagation of a point particle on M with a given metric g. We may ask how is the BRST charge Q is modified under a small variation of g, $g \to g + \delta g$. For this we first note that if $(M, g + \delta g)$ is diffeomorphic to (M, g), then the modified BRST charge should be equivalent to the original one. Thus, we expect that the deformation theory of Q should be a cohomology problem. In the example at hand, it is easy to see how this works. For a generic deformation of the metric we have

$$\delta Q = -c\delta g^{-1}(p, p)\,,$$

which is clearly Q-closed since c squares to zero. On the other hand, if $g+\delta g = f^*g$ for an infinitesimal diffeomorphism generated by a smooth vector field ξ, then $\delta Q = -\frac{1}{2}[Q, p(\xi)]$. Thus the nonequivalent deformations of the point particle action are given by the cohomology of the adjoint action of Q at degree 1, while the spectrum of physical states is isomorphic to the cohomology of Q at degree 0.

2.2 Scattering Matrix and Minimal Model

In particle physics, one is typically (though not exclusively) interested in transition amplitudes between ingoing and outgoing particles. In the absence of four-body interactions the three possibilities, s-, t-, and u-channel in physic parlance, for four particles to interact are given in Fig. 2.2. In the world-line formulation such an amplitude corresponding to the first graph in Fig. 2.2 is given by the generalization of (2.15), that is

$$\begin{aligned}&\int_{\tau_i}^{\tau_f} ds_2 \int_{\tau_i}^{s_2} ds_1 \int dc\, d\phi_f\, d\phi_i \\ &\quad \int D[\phi, b, c]\, e^{\,iI[\phi,b,c]}\, \bar{\psi}_4(\phi_f)\, \delta(c(s_2))\, \psi_3(\phi(s_2)) \\ &\quad \times \delta(c(s_1))\, \psi_2(\phi(s_1))\, \psi_1(\phi_i)\,.\end{aligned} \tag{2.23}$$

The insertions of the c-ghost delta functions are to remove the gauge fixing the proper times, s_1 and s_2, of the interaction vertices as before, since these are moduli to be integrated over. Expressed in terms of the evolution kernels, this becomes

$$\begin{aligned}&\int_{\tau_i}^{\tau_f} ds_2 \int_{\tau_i}^{s_2} ds_1\, d\phi_f\, d\phi_2 d\phi_1 d\phi_i\, \bar{\psi}_4(\phi_f)\, K(\tau_f, s_2, \phi_f, \phi_2) \\ &\quad \int dc\, c\, \psi_3(\phi_2)\, K(s_2, s_1, \phi_2, \phi_1) \int dc\, c\, \psi_2(\phi_1)\, K(s_1, \tau_i, \phi_1, \phi_i)\psi_1(\phi_i)\,.\end{aligned}$$

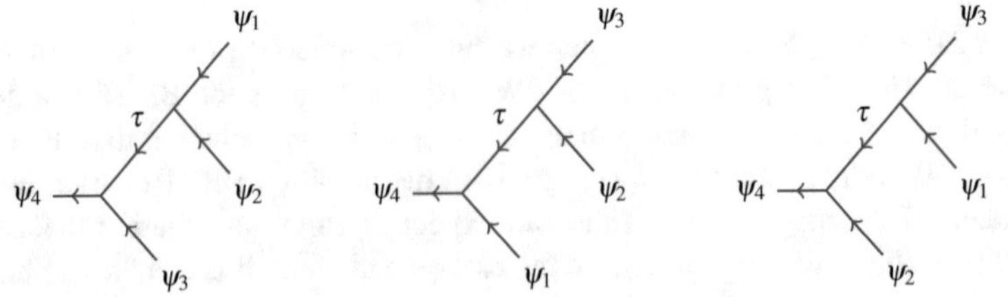

Fig. 2.2 The three distinct world-line diagrams for the scattering of four particles. The affine parameter τ parametrizes the "length" of the internal line

The first and the last insertions of the evolution kernel K correspond to the free propagation of the initial and final states. The standard procedure is to "amputate these external lines" in order to isolate the scattering process. The amplitude (2.23) then reduces to

$$\int_0^\infty ds\, d\phi_2 d\phi_1 \int dc\, (\bar{\psi}_4 * \psi_3)(\phi_2)\, K(s,\phi_2,\phi_1) \int dc\, (\psi_2 * \psi_1)(\phi_1)\,, \tag{2.24}$$

where $s = s_2 - s_1$. Here we used (2.12) as well as (2.15) and set $\tau_i = 0$ and $\tau_f = \infty$.

Remark 2.5 More precisely, the s-integral in (2.24) should be understood as follows. Starting from (2.23) one should consider τ_f of the form $\tau_f = ae^{i\theta}$ with $\theta \in (0, \frac{\pi}{2}]$ and preform the limit $a \to \infty$ in order to get (2.24). This will improve the convergence of the resulting s-integral for an open subset of possible momenta of the external states. The integrand is analytic in the first quadrant so that the result does not depend on the value of θ. To continue, we take $\theta = \frac{\pi}{2}$, which is equivalent to the substitution $\tau \to i\tau$, i.e., *Wick rotation* to imaginary time.

Returning to the canonical formalism, now with imaginary proper-time evolution operator $e^{-\mathcal{H}\tau}$, the expression (2.24) can be written equivalently as

$$\langle \psi_4 * \psi_3, Q^{-1}(\psi_1 * \psi_2)\rangle = \langle \psi_4, \psi_3 * Q^{-1}(\psi_1 * \psi_2)\rangle\,, \tag{2.25}$$

where we used the cyclicity ($\langle a * b, c\rangle = (-1)^{|b|}\langle a, b * c\rangle$) and where Q^{-1} is the propagator, the homotopy inverse of Q on the complement to H, $|b|$ is the degree of b. More precisely, let P be the projector to H, the cohomology of Q in V. Then

$$Q^{-1}Q + QQ^{-1} = \mathbb{1} - P\,, \qquad Q^{-1} = \frac{b}{\mathcal{H}}(\mathbb{1} - P)\,. \tag{2.26}$$

Note that due to (2.8) the action of the operator b is identified with $\int dc = \frac{\partial}{\partial c}$.

Remark 2.6 More precisely, in order to define Q^{-1} we first embed H in $\ker(Q)$

$$\ker(Q) = i(H) \oplus V_T\,, \tag{2.27}$$

where $V_T = \mathrm{Im}(Q)$ in V. If V_U is the linear complement to $\ker(Q)$ in V such that $\omega|_H$ and $\omega|_{V_T \oplus V_U}$ are symplectic and $\omega|_{V_T} = \omega|_{V_U} \equiv 0$, then Q is an isomorphism $V_U \cong V_T$. The homotopy inverse Q^{-1} is then defined as the inverse of Q on V_T and extended to be zero everywhere else.

The decomposition of any trivalent graph follows the same pattern. Thus, the pair Q^{-1} and $*$ together with the symplectic form are sufficient to describe the tree-level scattering of any number of spinless particles. We note that the moduli space of such labeled trees has no boundary since any two trees in Fig. 2.2 smoothly cross into each other when the affine parameter τ approaches 0. On the other hand, this decomposition is not unique. For instance, we could require that the affine parameter has to be larger than some value ϵ, say. This is equivalent to adding a piece of propagator of length ϵ to each leg of the cubic vertex. This new cubic vertex defines a product $l_2^\epsilon : V \otimes V \to V$, related to $*$ through

$$l_2^\epsilon(a, b) = e^{\mathscr{H}\epsilon}\left((e^{-\mathscr{H}\epsilon}\, a) * (e^{-\mathscr{H}\epsilon}\, b)\right) . \tag{2.28}$$

Note though that the composition of two binary products $l_2^\epsilon(l_2^\epsilon(a, b), d)$ still vanishes trivially due to the nilpotency of the ghost c. However, it is already clear from Fig. 2.2 that we need to introduce a four-vertex in order to reproduce the correct 4-point amplitude. Indeed, the zero-length propagator s-, t-, and u-channel trees do not touch anymore and we therefore need to add a quartic vertex

$$f_4^\epsilon(\psi_4, \psi_3, \psi_2, \psi_1) = \langle l_2^\epsilon(\psi_4, \psi_3), b\, e^{-2\epsilon\mathscr{H}}\, l_2^\epsilon(\psi_1, \psi_2)\rangle \,+\, (\text{perm.})\,.$$

Using again the identity

$$\begin{aligned}\omega(l_2^\epsilon(a, b), c) &= (-1)^{|a|+|b|+1}\langle l_2^\epsilon(a, b), c\rangle \\ &= (-1)^{|a|+1}\langle a, l_2^\epsilon(b, c)\rangle = -\omega(a, l_2^\epsilon(b, c))\,,\end{aligned}$$

we can rewrite the latter formula as

$$\begin{aligned} f_4(\psi_4, \psi_3, \psi_2, \psi_1) &= \omega(\psi_4, l_2^\epsilon(\psi_3, b_0\, e^{\epsilon\mathscr{H}}\, l_2^\epsilon(\psi_2, \psi_1))) \\ &\quad +\omega(\psi_4, l_2^\epsilon(\psi_1, b_0\, e^{\epsilon\mathscr{H}}\, l_2^\epsilon(\psi_2, \psi_3))) \\ &\quad +\omega(\psi_4, l_2^\epsilon(\psi_2, b_0\, e^{\epsilon\mathscr{H}}\, l_2^\epsilon(\psi_3, \psi_1))) \\ &\equiv \omega(\psi_4, l_3(\psi_3, \psi_2, \psi_1))\,. \end{aligned} \tag{2.29}$$

The triple product l_3 is trivially Q-closed,

$$[Q, l_3] \equiv Q \circ l_3 + l_3 \circ (Q \otimes \mathbb{1} \otimes \mathbb{1}) + l_3 \circ (\mathbb{1} \otimes Q \otimes \mathbb{1}) + l_3 \circ (\mathbb{1} \otimes \mathbb{1} \otimes Q) = 0\,, \tag{2.30}$$

again due to the nilpotency of c. As a consequence, the BV action for three-vertices with stubs is obtained simply by adding the quartic vertex to (2.16).

Of course, once a four-vertex is introduced this will imply a five-vertex upon substitution into a tree with 5 legs and so forth so that eventually we will end up with an infinite set of vertices, or maps $\{l_n^\epsilon\}$, with relations of the form given in (2.30).

Remark 2.7 It turns out that this procedure of adding stubs to the vertices has a familiar physical interpretation in terms of a tree-level Wilsonian effective action. Indeed, adding an infinitesimal stub of length ϵ to the cubic vertex is the same as inserting a momentum cut-off in the propagators since the values of $p^2 = g^{-1}(p, p) \gg \frac{1}{\epsilon}$ are exponentially suppressed. Thus, the contribution from large momenta to the scattering amplitudes are contained in the higher vertices $\{l_n^\epsilon, n = 3, 4, \ldots\}$. At order ϵ the effective action so obtained contains a four-vertex. At order ϵ^2 a quartic as well as a quintic vertex is produced. We can repeat this procedure until all momenta of the internal propagators are integrated out. The resulting effective action has no propagators left and therefore has the physical interpretation of a generating function for scattering amplitudes. We denote the corresponding maps by $\{s_n\}$.

The algebraic counterpart of this procedure, the *minimal model theorem*, which for the model at hand states that, given a dg-algebra $(V, *, Q)$, there exists a quasi-isomorphism[2] F from $(H, \{s_n\})$ to $(V, *, Q)$, which induces an isomorphism on the physical subspace contained in the cohomology H. Furthermore the maps $\{s_n\}$ are just the matrix elements of the scattering matrix of physical states.

There is a simple and intuitive procedure to obtain the minimal model map by constructing a perturbative solution to the equation of motion derived from the action (2.16). To begin with we split the field ψ into a "physical" state $P\psi = \psi^P \in H$ and its complement $\psi^U + \psi^T$ in $V_U \oplus V_T$. The subspace V_U corresponding to the projection map $P_U = -Q^{-1}Q$ represents the nonphysical states, i.e., the states not annihilated by Q, and the subspace V_T represents the space of trivial states, i.e., Q-exact states. We can then consistently set $\psi^T = 0$ and solve for ψ^U starting from the equation of motion for ψ. Since $Q\psi^P = 0$ by construction, we have

$$Q\psi^U = -\psi * \psi\,. \tag{2.31}$$

Acting on this equation with Q^{-1} we get

$$\begin{aligned}\psi^U &= -Q^{-1}\big((\psi^P + \psi^U) * (\psi^P + \psi^U)\big) \\ &= -Q^{-1}\big(\psi^P * \psi^P\big) + Q^{-1}\big(Q^{-1}\big(\psi^P * \psi^P\big) * \psi^P\big) \\ &\quad + Q^{-1}\big(\psi^P * Q^{-1}\big(\psi^P * \psi^P\big)\big)\,,\end{aligned} \tag{2.32}$$

up to terms of order four and higher. This gives a perturbative expression for ψ^U in terms of ψ^P which satisfies (2.31) up to an element in the kernel of Q^{-1}. Substituting the right-hand side of (2.32) into (2.31) and using (2.26) as well as $Q^2 = 0$ we find

$$P\big((\psi^P + \psi^U) * (\psi^P + \psi^U)\big) = 0\,, \tag{2.33}$$

[2]Cf. Remark 2.10 in Sect. 2.3 concerning the use of the term minimal model in the world of cyclic homotopy algebras.

which upon expanding ψ^U in terms of ψ^P using (2.32) produces an obstruction for ψ^P to give a solution to (2.32). This obstruction is of the form

$$\sum_{k\geq 2} s_k(\psi^P, \ldots, \psi^P) = 0\,, \tag{2.34}$$

which is a Maurer–Cartan equation for ψ^P. On the other hand, upon inspecting the first few terms in (2.33), that is,

$$\begin{aligned} s_2(\psi^P, \psi^P) &= P(\psi^P * \psi^P)\,, \\ s_3(\psi^P, \psi^P, \psi^P) &= -P\left[(Q^{-1}(\psi^P * \psi^P) * \psi^P) + (\psi^P * Q^{-1}(\psi^P * \psi^P))\right], \end{aligned}$$

and recalling that Q^{-1} is the propagator, we see that

$$\langle \psi^P, s_2(\psi^P, \psi^P)\rangle$$

is the scattering matrix for a three-particle scattering (which usually vanishes for the kinematical reasons), while

$$\langle \psi^P, s_3(\psi^P, \psi^P, \psi^P)\rangle$$

is the scattering matrix for a four-particle scattering. It is not hard to see that this identification holds for an arbitrary power of ψ^P. To summarize, the maps s_n which are defined on the cohomology H can be interpreted as the scattering matrices for an $n+1$ particle scatterings. Equation (2.32)[3] induces a nonlinear map $F : H \to V$. Explicitly, let SV be the symmetric algebra

$$SV = \bigoplus_{n=0}^{\infty} (V^{\otimes n})_{\text{sym}},$$

where $V^{\otimes 0} = \mathbb{C}$ and similarly, SH the symmetric algebra on H. Then we can write

$$\begin{aligned} \psi = F(\psi^P) = \psi^P + \psi^U &= \pi_1(\mathbb{1} - Q^{-1}*)^{-1}(e^{\psi^P} - 1) \\ &= \psi^P - Q^{-1}(\psi^P * \psi^P) + Q^{-1}(Q^{-1}(\psi^P * \psi^P) * \psi^P) + \cdots\,, \end{aligned}$$

where $\pi_1 : SV \to V$ denotes the projection on one output. Similarly, we have

$$s_n = P\pi_1 * (\mathbb{1} - Q^{-1}*)^{-1} i_n\,, \tag{2.35}$$

[3]Here we are somewhat cavalier about the proper notion of embedding of ψ^P in V.

where $i_n : (H^{\otimes n})_{\text{sym}} \to SH$ is the inclusion map. It is then not hard to see that F is a chain map, that is,

$$(Q + *)F = FS\,, \tag{2.36}$$

where $S = \sum_{n=0}^{\infty} s_n$. Furthermore, $(Q+*)^2 = 0$ due to (2.17) and similarly $S^2 = 0$. We will use the same symbol for the induced morphism of the underlying algebraic structures, i.e., we will write $F\ : (H, \{s_n\}) \to (V, Q, *)$. Let us note that F also respects the cyclic structure, i.e.,

$$F^*\omega = \omega_H\,, \tag{2.37}$$

with F^* denoting the pullback of F. In this sense, we actually have an induced morphism $F\ : (H, \{s_n\}, \omega_H) \to (V, Q, *, \omega)$ of the underlying algebraic structures including also the respective odd symplectic forms.

Remark 2.8 We can also express the minimal model theorem directly in terms of the BV action. We take the BV vector field $Q = \{S, -\}$ in (2.18) and expand it in the formal neighborhood $[\phi_0]$ around a point ϕ_0 in the degree 0 subspace of the space of fields $\mathscr{F}$

$$Q \overset{form}{=} Q^{(0)}_{\phi_0} + Q^{(1)}_{\phi_0} + Q^{(2)}_{\phi_0} + \cdots\,.$$

If ϕ_0 is an element of the Euler–Lagrange subspace $\mathscr{EL}$ (i.e., the space of classical solutions) of $\mathscr{F}$, then $Q^{(0)}_{\phi_0}$ vanishes and the set $\{Q^{(k)}_{\phi_0}, k > 0\}$ defines degree one vector fields on $T_{\phi_0}\mathscr{F}$. In our example with $\phi_0 = 0$,

$$\begin{aligned} Q^{(1)}_0 &= \omega^{ij}\phi^k\omega(e_k, Qe_i)\partial_{\phi_j}\,,\\ Q^{(2)}_0 &= \omega^{ij}\phi^k\phi^r\omega(e_k, e_r * e_i)\partial_{\phi_j}\,. \end{aligned}$$

More generally, if we restrict the neighborhood $[\phi_0]$ to be in $\mathscr{EL}/Q$, then the $\{Q^{(k)}_{\phi_0}\}$ define linear maps

$$Q^{(k)}_{\phi_0} = \phi^{i_1}\cdots\phi^{i_k} l^i{}_{i_1\cdots i_k}\partial_{\phi_i}\ : T_{\phi_0}\mathscr{F} \to S(T_{\phi_0}\mathscr{F})\,.$$

These maps are dual to the multilinear maps $\{l_n\}$ defined above, that is,

$$l_n(e_{i_1}, \ldots e_{i_n}) = l^i{}_{i_1\cdots i_n}\, e_i\,.$$

Furthermore, the collection $\{Q^{(k)}_{\phi_0}\}$ induces the Maurer–Cartan equation (2.34) on H.

Remark 2.9 One might worry that, since the action (2.16) was constructed such as to reproduce an n-particle scattering in a specific background, the whole construction may not be background independent . That this is not so follows from the observation that the definition of Q in (2.10) is universal, i.e., independent of ψ. In fact, the presentation of the ψ^3-field theory in this section is "reversed." Indeed, the usual starting point is the action (2.16), which is background independent and one then derives the scattering amplitudes (2.15). However, we chose this presentation for pedagogical reasons as will hopefully become clear in sections on string field theory.

Although we have elaborated here on the simple example of a point particle with the trivial ghost and gauge sectors, most of the algebraic structure discussed in this section will hold for a generic BV theory and in particular for string field theory. One important modification is that, due to the non-trivial gauge structure of the string, the nilpotent algebras we encountered for the point particles will be replaced by homotopy Lie- or associative algebras which will be the relevant algebraic structure for the string.

2.3 Summary, Comments, and Remarks Towards Part II

The path integral (2.11) has an interpretation as a symmetric monoidal functor E from the one-dimensional bordism category to the category of vector spaces. This functor associates to a point $*$ the Hilbert space $E(*) = V$ of functions on the "mapping" space $\{* \to M \times \mathbb{R}^{0|1}\} = M \times \mathbb{R}^{0|1}$, where $\mathbb{R}^{0|1}$ is the odd line generated by the ghost c. The functional integral associates to a bordism (interval $\Delta\tau$) between two points a map $V \to V$ through the integration of the kernel $K(\Delta\tau, \phi_1, \phi_0)$ against the wave functions $\psi(\phi_0, c)$. We will not elaborate on this point of view any further.

The point we wished to elucidate here is that the space V naturally carries the structure of a nilpotent abelian differential graded (dg) algebra $(V, *, Q)$ equipped with an invariant odd symplectic form ω. In addition,

$$S[\psi] = \frac{1}{2}\omega(\psi, Q\,\psi) + \frac{1}{3}\omega(\psi, \psi * \psi) \tag{2.38}$$

defines a BV action on V, i.e., it satisfies the classical BV master equation $\{S, S\} = 0$. Equivalently, $(Q + *)^2 = 0$. Furthermore, with $\psi = f_0 + f_1 c$, f_0 of degree 0 and f_1 of degree -1, we have

$$\psi * \psi = c f_0 f_0\,, \quad Q\psi = 1/2c(\Box + m^2) f_0 \text{ and } \langle \psi, \psi' \rangle = \int_M f_0 f_1' + f_0' f_1\,.$$

In particular, f_1 is the anti-field corresponding to f_0. Finally, we see that

$$S = \int_M \frac{1}{2} f_0(\Box + m^2) f_0 + \frac{1}{3} f_0^3 .$$

Not surprisingly, what we described in a rather involved way is the ordinary scalar field theory with cubic interaction and no gauge symmetry. The action S satisfies BV master equation trivially, there are no anti-fields in the action.

Nevertheless, it may prove useful to extend an ordinary field theory to a BV one trivially just adding the anti-fields as above. Many concepts useful in quantum field theory, e.g., Schwinger–Dyson equations, Ward identities, etc., have their conceptual origin in BV formalism.

As already mentioned in the introduction, classical BV actions arise naturally from representations of the cobar construction on cyclic operads. The n-valent interaction vertices in the BV action correspond to $(n - 1)$-ary degree one operations (brackets, products, etc.) of the corresponding homotopy algebras. This will thoroughly be described in Part II. Concerning our case of the point particle, the relevant operad is the cyclic commutative operad. The corresponding cyclic L_∞-algebra has all higher brackets trivial, the binary (graded *symmetric* degree one) bracket is the one given by $*$, which is just the result of the ordinary (graded) commutative point-wise product multiplied by the ghost c. Similarly, in a trivially BV extended quantum field theory, all higher brackets will be given by the powers of the ordinary product $c\psi^n$, which can be seen as generated by the binary one, see, for example, (2.29).

The physical scattering amplitudes of n identical particles are evaluated by drawing all inequivalent labeled and directed trees with n inputs and one output as in Fig. 2.2 constructed from the cubic vertex which defines the commutative product of differential graded algebra $(V, *, Q)$ underlying the action functional (2.38). Two trees are equivalent if they are obtained by permutation of two labels at the same vertex. Each internal line between two vertices represents the insertion of a "propagator" $\frac{1}{i}Q^{-1}$. The physical amplitude is then given by evaluation with the symplectic form. For instance, for the scattering $(1, 2, 3 \to 4)$ the trees are given in Fig. 2.2 with the resulting physical amplitude

$$\frac{1}{i}\omega(\psi_4, \psi_3 * Q^{-1}(\psi_1 * \psi_2)) + \frac{1}{i}\omega(\psi_4, \psi_1 * Q^{-1}(\psi_3 * \psi_2)) \\ + \frac{1}{i}\omega(\psi_4, \psi_2 * Q^{-1}(\psi_3 * \psi_1)) .$$

This construction of tree-level scattering amplitudes for the relativistic scalar particle is the physical equivalent of the construction of the minimal model on the Q-cohomology H for cyclic L_∞-algebras.

Remark 2.10 To give, in our setting of cyclic homotopy algebras, a sensible meaning to the term "minimal model," we would like to interpret the map (2.35) as a morphism from the cyclic L_∞-algebra $(H, \omega|_H, S)$ on the cohomology H to the original L_∞-algebra (V, ω, L). Formulas (2.36) and (2.37) suggest one possible definition of a morphism in a category of cyclic L_∞-algebras as a map (in general a nonlinear one) between the underlying vector spaces compatible with the respective cyclic L_∞-structures. Obviously, a map respecting symplectic structures might seem to be a too restrictive one, the dimension of the source vector space cannot be bigger than that of the target vector space. A possible way out would be to consider Lagrangian correspondences instead of maps. We will not need this generalization. Hence, we will not comment on this any further.

The minimal model defined on the subspace of physical states is then quasi-isomorphic to the L_∞-algebra defined by the original action S on the space of all states. Formula (2.35) for the tree-level scattering amplitudes/brackets of the minimal model is the content of the homological perturbation lemma. It expresses the change of the trivial differential on functions on the Q-cohomology H induced by the perturbation of the BRST operator Q by the product $*$ on functions on V.

Finally, we would like to comment on the relation to the operadic formulation of cyclic L_∞-algebras in Part II of this book. The maps l_n or equivalently, their duals $Q^{(k)}_{\phi_0}$ and indeed, the classical BV action S, define a cyclic L_∞-algebra. This is an algebra over the cobar construction of the cyclic commutative operad briefly recalled at the beginning of Sect. 7.2 in Part II. The minimal model construction, when expressed by formula (2.35), applies directly to any algebra over the cobar construction of a general cyclic operad and even more generally to any algebra over the Feynman transform of a general modular operad. In the form presented here it also applies to representations of cobar constructions of properads and hence, in particular, to IBL_∞-algebras.

Further Reading

An intuitive pedagogical account of the quantization of the relativistic point particle can be found in:

- A. M. Polyakov, "Gauge fields and strings", Harwood, 1987.

A standard reference for BRST and BV quantization is:

- M. Henneaux and C. Teitelboim, "Quantization of gauge systems", Princeton University Press, Princeton 1992.

Other useful references include:

- S. Weinberg, "The quantum theory of fields Vol. II", Cambridge University Press, 2005,

- J. Gomis, J. Paris and S. Samuel, "Antibracket, antifields and gauge-theory quantization, Physics Reports", Volume 259, 1995, and
- M. Alexandrov, A. Schwarz, O. Zaboronsky and M. Kontsevich, "The geometry of the master equation and topological quantum field theory", Int. J. Mod. Phys. A 12, 1405 (1997),

or, for a more mathematically minded reader,

- A. Cattaneo and N. Moshayedi, "Introduction to the BV-BFV formalism", arXiv:1905.08047.

Although we do not pursue this avenue in this book, it is possible to formulate Yang-Mills theory and gravity on the world-line with (extended) world-line supersymmetry, see:

- P. Dai, Y. Huang and W. Siegel, "Worldgraph approach to Yang-Mills amplitudes from $N = 2$ spinning particle", JHEP 0810 (2008) 027

for Yang-Mills theory, and

- R. Bonezzi, A. Meyer and I. Sachs, "Einstein gravity from the $N = 4$ spinning particle", JHEP 1810, 025 (2018)

for gravity. There are many textbooks explaining the diagrammatic evaluation of scattering amplitudes. A popular reference for physicists is:

- M.E. Peskin and D.V. Schroeder, "An introduction to quantum field theory", Westview Press, 2015.

The minimal model construction as a direct application of the homological perturbation lemma is discussed in:

- M. Doubek, B. Jurčo and J. Pulmann, "Quantum L_∞-algebras and the homological perturbation lemma", Commun. Math. Phys. 367(1), (2019) 215–240.

3 String Theory

A natural 1-dimensional generalization of the point particle is its blow-up into an open or closed string, cf. Fig. 3.1. It turns out that both possibilities are meaningful and, in fact, at quantum level the former implies the latter. The action for the open string turns out to be structurally very similar to that of the point particle described in the previous chapter. Indeed, one can show that the action (2.16) with suitably defined $(\omega, Q, *)$ corresponds to a decomposition of the moduli space of bordered Riemann surfaces with punctures on the boundary. For the closed string, however, this is not the case and an infinite number of higher order vertices has to be added to (2.16). In this chapter we will describe some features of string theory relevant for the rest of this book. For more details we refer to the original literature listed at the end of the chapter.

3.1 Closed Strings

Let us start with a one-dimensional generalization of the world-line action (2.1)

$$I[\phi, h] = \frac{1}{4\pi\alpha'} \int_{\Sigma} g(d\phi, \wedge^* d\phi)\,, \tag{3.1}$$

where now $\phi : \Sigma \to M$, while Σ is topologically a cylinder equipped with a pseudo-Riemannian world-sheet metric. However, in order to have a well-defined measure on the space of world-sheets we consider the Wick-rotated, or Riemannian world-sheet metric h. This is in analogy with the discussion after (2.24). Furthermore, $1/\alpha'$ is the string tension which sets the unit for the masses of the excitations of the string. In what follows, we work in units where $\alpha' = 1$ since we are not interested in the particle limit ($\alpha' = 0$), nor the tensionless limit ($\alpha' = \infty$) where all excitations are massless.

© Springer Nature Switzerland AG 2020

M. Doubek et al., *Algebraic Structure of String Field Theory*, Lecture Notes in Physics 973, https://doi.org/10.1007/978-3-030-53056-3_3

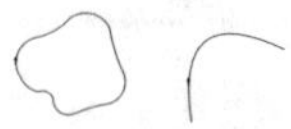

Fig. 3.1 Closed and open string

The constraint equation obtained upon varying this action with respect to h now reads $T^{\phi}_{ab} = 0$, where

$$T^{\phi}_{ab} = -\frac{4\pi}{\sqrt{h}} \frac{\delta I[\phi, h]}{\delta h^{ab}} \tag{3.2}$$

is a symmetric and traceless world-sheet (stress) tensor.

Remark 3.1 The world-sheet stress tensor is traceless, since the diagonal part of h does not appear in S. However, the absence of anomalies for this "Weyl invariance" implies strong conditions on the manifold M. We will assume that M is of dimension 26 equipped with a flat metric to avoid this complication.

In analogy with the point particle, the Faddeev–Popov procedure takes care of the redundancy, due to the world-sheet diffeomorphism invariance of (3.1) generated by the vector field ξ,

$$\delta\phi = -[\xi, \phi]\,, \qquad \delta h = -\mathcal{L}_{\xi} h\,.$$

The resulting gauge-fixed Euclidean evolution kernel is then given by

$$Z(h, \phi_f, \phi_i) = \int_{\phi_{\partial\Sigma_i} = \phi_i}^{\phi_{\partial\Sigma_f} = \phi_f} D[\phi, b, c]\, e^{-I[\phi, b, c, h]}\,, \tag{3.3}$$

where $\phi_{f/i}$ are maps from the boundary components $\partial\Sigma_{f/i}$ to M and

$$I[\phi, b, c, h] = \frac{1}{4\pi}\int_{\Sigma} g(d\phi, \wedge^* d\phi) - \frac{i}{2\pi}\int_{\Sigma} \sqrt{h}\, b_r{}^s\, \nabla_s c^r\, d\tau d\sigma\,. \tag{3.4}$$

Here ∇ is the covariant derivative compatible with the metric h on the world-sheet. Geometrically, c is an odd vector field on Σ while b is an odd, symmetric traceless tensor. The residual BRST symmetry is (with $\xi \to -ic$)

$$\delta_{BRST}\phi = i[c, \phi]\,, \quad \delta_{BRST} b = iT\,, \quad \delta_{BRST} c = \frac{i}{2}\mathcal{L}_c c\,, \tag{3.5}$$

where $T = T^{\phi} + T^{g}$, with T^{ϕ} given in (3.2) and $T^{g} = \mathcal{L}_c b$ is the stress tensor of the ghost sector. The extra factor of i missing in (2.7) is due to the Wick-rotation discussed above. This is the appropriate one-dimensional generalization of (2.6) and (2.7).

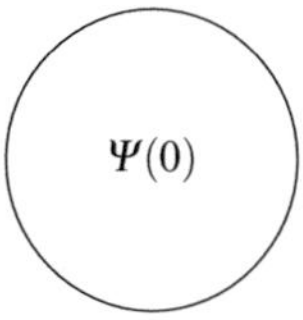

Fig. 3.2 The unit disk with $\psi(0)$ inserted at the origin where, more generally, $\psi(P) \equiv \psi(\phi(P), \partial_z\phi(P), \ldots)$

To complete the construction, we should provide a generalization of the vector space V. Since the path integral measure in (3.3) is invariant under conformal mappings that map $\tau = -\infty$ to the origin of the complex plane, we can define a vector space spanned by polynomials in the fields and their derivatives $\psi(\phi, b, c, \partial_z\phi, \ldots)$ inserted at the origin of the complex plane. A convenient way to parametrize the vector ψ goes as follows. Let $\{\bar{\phi}, \bar{b}, \bar{c}\} : \partial D \to M$ be the restriction of the maps $\{\phi, b, c\}$ to the boundary of the unit disk. We then evaluate the path integral measure in (3.3) on the unit disk with Euclidean metric $ds^2 = dzd\bar{z}$ subject to the boundary condition $\{\bar{\phi}, \bar{b}, \bar{c}\}$ and $\psi(\phi, b, c, \partial_z\phi, \ldots)$ inserted at the origin of the disk, cf. Fig. 3.2. Concretely,

$$\psi(\bar{\phi}, \bar{b}, \bar{c}) = \int_{\{\phi,b,c\}|_{\partial D}=\{\bar{\phi},\bar{b},\bar{c}\}} D[\phi, b, c]\, e^{-I(\phi,b,c,h)}\psi(0)\,. \tag{3.6}$$

It turns out that there is some redundancy in this representation due to the invariance under reparametrizations of the circle $|z| = 1$. These are generated by the vector fields $\xi_n = e^{in\theta}\partial_\theta$ that generate the Lie algebra of Diff(S^1). Consequently, the vectors $\{\psi\}$ should transform in a representation of Diff(S^1), i.e. they transform as tensors of homogeneous degrees. This important structural difference against the point particle described in the previous chapter gives rise to an infinite dimensional gauge redundancy in string theory. The vector fields ξ_n can be continued as mereomorphic vector fields inside the disk as

$$i\xi_n = z^{1+n}\partial_z - \bar{z}^{1-n}\partial_{\bar{z}} \equiv L_n - \bar{L}_{-n}\,.$$

The vector fields L_n so defined realize the familiar Witt or Virasoro algebra with non-vanishing commutation relations

$$[L_n, L_m] = (n-m)L_{n+m}\,, \qquad [\bar{L}_n, \bar{L}_m] = (n-m)\bar{L}_{n+m}\,, \qquad n \in \mathbb{Z}\,.$$

This is just a double copy of the Lie algebra of Diff(S^1). We assume the central extension to be absent due to anomaly cancelation in 26 dimensions (see Remark 3.1).

In the operator formalism one starts with a classical solution (critical points of (3.4)) given by mereomorphic functions with an expansion

$$i\partial_z\phi^\mu(z) = \sum_{n\in\mathbb{Z}} \frac{\alpha_n^\mu}{z^{n+1}}\,, \quad c(z) = \sum_{n\in\mathbb{Z}} \frac{c_n}{z^{n-1}}\,, \quad b(z) = \sum_{n\in\mathbb{Z}} \frac{b_n}{z^{n+2}}$$

$$i\partial_{\bar z}\phi^\mu(\bar z) = \sum_{n\in\mathbb{Z}} \frac{\tilde\alpha_n^\mu}{\bar z^{n+1}}\,, \quad \tilde c(\bar z) = \sum_{n\in\mathbb{Z}} \frac{\tilde c_n}{\bar z^{n-1}}\,, \quad \tilde b(\bar z) = \sum_{n\in\mathbb{Z}} \frac{\tilde b_n}{\bar z^{n+2}}\,.$$

where $\mu = 1, \ldots, 26$ refers to the Cartesian coordinates on $\mathbb{R}^{26}$. To continue, one can either postulate the algebra

$$[\alpha_n^\mu, \alpha_m^\nu] = n g^{\mu\nu}\delta_{n,-m}\,, \quad [b_n, c_m] = \delta_{n,-m}\,,$$

where $g^{\mu\nu}$ are the components of the inverse of the metric tensor g on $\mathbb{R}^{26}$ and $[-,-]$ denotes the graded commutator or, equivalently, start from the operator product expansion

$$i\partial\phi(z) i\partial\phi(w) \simeq \frac{\mathbb{1}}{(z-w)^2} + O((z-w)^0)\,, \quad b(z)c(w) \simeq \frac{\mathbb{1}}{(z-w)} + O((z-w)^0)$$

to derive this algebra. In analogy with the point particle, we assign ghost number one to c and $\tilde c$, minus one to b and $\tilde b$ and zero to ϕ. For the point particle we identified the ghost number with the degree, whereas for the string we define the degree as deg $\equiv$ ghost number-2 for consistency with the standard convention in the BV formalism. In the operator formalism ∂_z and $\partial_{\bar z}$ are represented by $T(z)$ and $\bar T(\bar z)$, respectively, that is,

$$L_n = \oint \frac{dz}{2\pi} z^{n+1} T(z)$$

and analogously for $\bar L_n$.

With this V acquires the structure of a Fock space where the vacuum state $|0, \mathbf{k}>$ corresponds to the insertion of $\psi = e^{ik_\mu\phi^\mu}$ at the origin of D,

$$\alpha_n^\mu |0, \mathbf{k}> = \oint \frac{dz}{2\pi} z^n \partial_z\phi^\mu(z)\psi(0)$$

and analogously for $\partial_{\bar z}\phi$, b, $\tilde b$, c, and $\tilde c$. In particular, $\alpha_n^\mu|0, \mathbf{k}> = 0$ for $n > 0$. There is a canonical inner product on V, the *BPZ* inner product, obtained through the gluing of two disks along the boundary at $|z| = 1$ with the opposite orientation, that is

$$\langle \psi_1, \psi_2 \rangle := \int D[\phi, b, c] \bar{\psi}_1[\phi, b, c] \psi[\phi, b, c]$$
$$= \lim_{|z| \to 0} \left\langle (I^* \psi_1)(z, \bar{z}) \psi_2(z, \bar{z}) \right\rangle , \tag{3.7}$$

where $I(z, \bar{z}) = (1/z, 1/\bar{z})$ and $I^* \psi$ is the pullback of ψ.

Next we turn to the BRST charge Q. Given the algebraic structure described above, it is easy to show that the BRST transformations (3.5) are generated by

$$Q = \oint \frac{dz}{2\pi i} c(z) \left(T^\phi(z) + \frac{1}{2} T^g(z) \right) + \oint \frac{d\bar{z}}{2\pi i} \tilde{c}(\bar{z}) \left(\bar{T}^\phi(\bar{z}) + \frac{1}{2} \bar{T}^g(\bar{z}) \right) .$$

The mode expansion of Q is of the form

$$Q = c_0^+ L_0^+ + c_0^- L_0^- + \cdots , \qquad c_0^\pm = \frac{1}{2}(c_0 \pm \tilde{c}_0) .$$

Proceeding as in (2.9), we choose the ghost vacuum so that

$$b_0^- \psi(\phi, b, c, \partial_z \phi, \cdots) = 0.$$

The physical subspace V_{phys} should then be given by the semi-relative cohomology $\mathrm{coh}(Q, b_0^-)$ (i.e., by the cohomology restricted to the kernel of b_0^-) at ghost number 2. In the Siegel gauge , $b_0^+ \psi = 0$, V_{phys} is then represented by the set of physical states found in the standard textbook presentation.

In order to ensure unitarity, that is the positivity of the inner product on the cohomology, it is necessary that pure gauge, i.e. a Q-exact state, has a vanishing inner product (3.7) with physical states. This amounts to the condition

$$L_n \psi(\phi, b, c, \partial_z \phi, \cdots) = \bar{L}_n \psi(\phi, b, c, \partial_z \phi, \cdots) = 0 , \quad n \geq 0 . \tag{3.8}$$

Algebraically, (3.8) states that $\psi(z, \bar{z})$ is a conformal primary field of dimension zero,

$$T(z)\psi(w, \bar{w}) \simeq \frac{1}{(z-w)^2}\psi(w, \bar{w}) + \frac{1}{z-w}\partial_w \psi(w, \bar{w}) + O((z-w)^0)$$

and analogously for $\bar{T}(\bar{z})$. Geometrically, the on-shell condition expresses the fact that physical states $\psi(\phi)$ are invariant under holomorphic reparametrizations of the disk that preserve the origin. Generic states will not be conformal primaries but they should still be invariant under rotations of the disk which is equivalent to

$$(L_0 - \bar{L}_0)\psi(\phi, b, c, \partial_z \phi, \cdots) = 0.$$

To understand the geometric origin of the ghost number 2 condition, we return to the inner product (3.7). A more careful inspection of (3.7) reveals that the correlation function $\langle \dots \rangle$ is zero unless we saturate it with three c-ghost and three $\tilde{c}$-ghost insertions, in other words, the correlator $\langle \dots \rangle$ has ghost number -6. These ghost zero modes correspond to the vector fields that generate the action of the automorphism group $SL(2,\mathbb{C})$ of S^2. Since they leave the reference metric $dzd\bar{z}$ invariant, they play no role in the gauge fixing procedure. This is why they show up as zero modes of the ghost action. On the other hand, the punctures of the two ψ-insertions are not invariant. One way to take care of this is to attach each insertion of ψ to a $c\bar{c}$-pair. This is equivalent to restricting gauge fixing to diffeomorphisms that preserve the insertion point. Thus, an equivalence class ψ of physical states has a representative of the form

$$\psi(0) = \psi(\phi(0), \partial_z\phi(0), \dots)c(0)\bar{c}(0)\,.$$

This takes care of four ghost zero modes in (3.7). Nonphysical states, on the other hand, do not correspond to conformal primaries, as we have already mentioned, therefore they depend not just on the position of the operator, but also on the holomorphic reparametrization of the disk. Consequently, such states may involve arbitrary polynomials of derivatives of the ghost fields.

The two remaining zero modes correspond to rotation and scaling of the unit disc, respectively. We have encountered the latter already for the point particle and it corresponds to time translation generated by Q, whereas the former is simply redundant due to the constraint $L_0^- = 0$. Furthermore, using $b_0^\dagger = b_0$, one easily checks that the inner product $\langle \psi_1, \psi_2\rangle$ is degenerate on states containing b_0^-. To remedy this using $[b_0^-, c_0^-] = 1$, we redefine the inner product as

$$\Omega(\psi_1, \psi_2) := \langle \psi_1, c_0^- \psi_2\rangle\,.$$

Due to the c_0^--insertion, Ω is graded anti-symmetric,

$$\Omega(A, B) = (-1)^{(|A|+1)(|B|+1)}\Omega(B, A)\,.$$

Furthermore, $\Omega(QA, B) = (-1)^{|A|}\Omega(A, QB)$. Now we have all necessary tools to write down the quadratic action for the closed string field. It is simply

$$S[\psi] = \frac{1}{2}\Omega(\psi, Q\psi)\,. \tag{3.9}$$

Just like the point particle, the string has a propagator, that is a homotopy inverse for Q, which we denote by Q^{-1}. In order to define it, repeating the steps in Remark 2.6, we fix the gauge, which amounts to fixing a representative for every element of the cohomology H. More precisely, the gauge fixing determines a map

$$i : H \to V,$$

which sends an element of the cohomology to its representative. We will call i the inclusion map. We also have the projection

$$\pi : V \to H.$$

Obviously, the map $P := i \circ \pi : V \to V$ satisfies $P^2 = P$ and the image V_P of P is isomorphic to H. This means that V_P represents the physical states. Moreover P is a chain map, i.e. $PQ = QP = 0$ and it induces the identity map on cohomology. This implies that P is homotopic to $\mathbb{1}$, i.e. there is a map $Q^{-1} : V \to V$ such that

$$P - \mathbb{1} = Q^{-1}Q + QQ^{-1}.$$

Note that $P^2 = P$ implies $(Q^{-1})^2 = 0$. Physically we can identify Q^{-1} as the propagator corresponding to the chosen gauge. We demand $Q^{-1}P = PQ^{-1} = 0$, which means that we set the propagator to be zero on the space of physical states. The subspace V_U corresponding to the projection map $P_U = -Q^{-1}Q$ represents the nonphysical states, i.e. the states not annihilated by Q, and the subspace V_T represents the space of trivial states, i.e. Q-exact states. To summarize, choosing a gauge in SFT determines a harmonious Hodge decomposition (compatible with the odd symplectic form), which decomposes the state space into physical, nonphysical, and trivial states. In Siegel gauge $b_0^+\psi = 0$ we have

$$Q^{-1} = \frac{b_0^+}{L_0^+}(\mathbb{1} - P).$$

3.2 Interactions

In order to work out the higher order corrections to the string action (3.9), we could proceed, as for the point particle in Chap. 2, by considering the propagation of a string in the background field corresponding to some external state ψ. In this way one obtains a prescription for calculating scattering matrix elements. However, from the point of view of string field theory, it is easier to just consider the geometrical vertex that describes the joining of two strings as in Fig. 3.3.

If we remove the free propagation of the external strings, we are left with a 3-punctured sphere with coordinate curves homotopic to the punctures, where the external states can be inserted by identifying the boundary of the unit disk of Fig. 3.2 with the coordinate curve. There is a phase ambiguity which corresponds to the

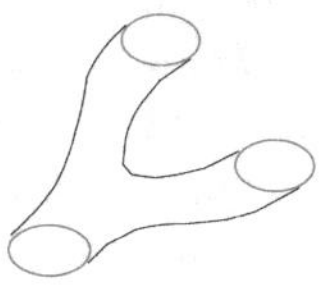

Fig. 3.3 Merging of two strings

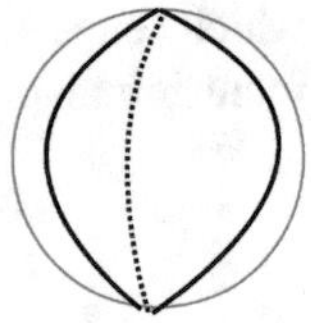

Fig. 3.4 Contact vertex for three strings

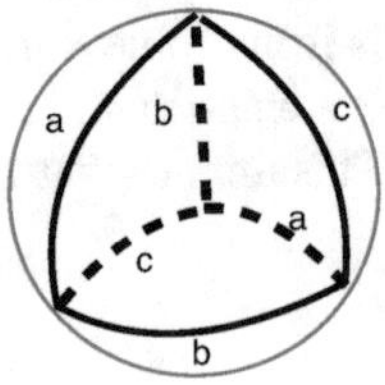

Fig. 3.5 A tetrahedron describing the scattering of four strings

rotation of these curves around the puncture. Due to the restriction $b_0^- \psi = L_0^- \psi = 0$, $\psi \in V$, this prescription is well defined.

Let us first consider a contact interaction for 3 strings. Without loss of generality, we can assume that each of the three external strings has circumference 2π. The 3-string vertex, denoted by ν_3, can be represented by a 2-sphere with three semicircles from the north to the south pole at the relative angle of $2\pi/3$ as in Fig. 3.4.

This defines a symmetric product on V upon gluing of the three semicircles with the boundaries of the disks used for representing the respective states as in (3.6). For the point particle, this vertex already provides a consistent action as we saw in Chap. 2. However, for the string that cannot be so as can be seen by examining the four-point vertex represented by a spherical tetrahedron in Fig. 3.5 subject to the condition that the boundary of each face has length 2π (four equations).

Since the tetrahedron has six edges, we are left with two variables which in turn parametrize the moduli space $\hat{\mathscr{P}}_4$ of 4-punctured spheres together with the corresponding coordinate curves around each puncture. The solution to the four constraint equations is best represented by a tetrahedron with opposite edges of equal length a, b, and c, respectively, see Fig. 3.5, subject to the constraint

$$a + b + c = 2\pi \,. \tag{3.10}$$

Solutions to this constraint give a parameterization of the relevant moduli space for four punctures. It is easy to see that not all of the moduli space can be covered by joining two cubic vertices with a closed string propagator represented by an internal cylinder of circumference 2π. Indeed, the tetrahedron obtained in this way, in the limit of a collapsing propagator, has two edges with length π, see Fig. 3.6.

The result is that by joining two cubic vertices in all possible, nonequivalent ways with a propagator of positive or vanishing length, one covers the part of the moduli space of the 4-vertex satisfying (3.10), which is the complement to the subset parametrized by (3.10) subject to the condition $a, b, c < \pi$. Gluing two cubic vertices in all possible, nonequivalent ways gives the boundary of that region. In order to cover the whole moduli space one needs to add an elementary four-vertex

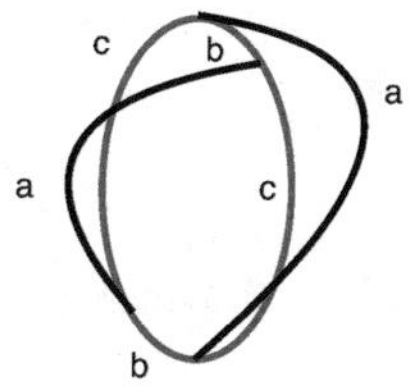

Fig. 3.6 A typical 4-point vertex obtained by joining two cubic vertices with a zero-length propagator. The edge a has length π

ν_4 as in Fig. 3.5 with $a, b, c < \pi$. This result can be cast into the form of a geometric BV equation

$$\partial \nu_4 + \frac{1}{2}\{\nu_3, \nu_3\} = 0\,, \tag{3.11}$$

which expresses the fact that, after a suitable compactification, the moduli space for four punctures built from vertices and propagators has no boundary. Here ∂ is the boundary operator, that is, $\partial \nu_4$ is the restriction of ν_4 to the subset where one of the edges has length π. The bracket $\{-, -\}$ stands for the twist-sewing of punctures of the two cubic vertices by identifying the local coordinate z around the first puncture with that around the second puncture through $z' = I(z) = 1/z$.

The twist-sewing comes about as follows. There is an ambiguity of determining local coordinates of coordinate curves parametrized by an angle $\theta \in [0, 2\pi)$, representing all possible rotations. Thus, the sewing of punctures with prescribed coordinate curves naturally generates a 1-parameter family of Riemann surfaces associated with the twist angle θ in local coordinates. The resulting vertex is subsequently symmetrized with respect to the remaining punctures of the combined surface. Of course, this will not stop at the four vertex. For the same reason there will be a quintic vertex, etc. In contrast to the point particle, closed string field theory is necessarily non-polynomial, with only the first few of these vertices known explicitly.

We will describe the geometric structure of this decomposition in more detail below but before that we discuss how to dress this geometric structure with physical states. For the cubic vertex in Fig. 3.4 this is done, as mentioned above, by identifying the round disk used for representing the state in (3.6) with the faces of the cubic vertex through a conformal mapping in such a way that the origin is mapped to the puncture and the boundary of the disk is mapped to the two edges of the face. In this mapping there is a one parameter ambiguity corresponding to a rotation of the boundary but, as we saw, this has the trivial action on V. In this way we obtain a cubic field theory vertex $\hat{f}_3$ and a bilinear, graded symmetric map $\hat{l}_2 : V \otimes V \to V$ through

$$\Omega(\psi_1, \hat{l}_2(\psi_2, \psi_3)) := \hat{f}_3(\psi_1, \psi_2, \psi_3)\,.$$

Similarly, for the elementary four-vertex one has

$$\Omega(\psi_1, \hat{l}_3(\psi_2, \psi_3, \psi_3)) := \hat{f}_4(\psi_1, \psi_2, \psi_3, \psi_4),$$

where the maps $\hat{l}_n : V \otimes V \otimes \cdots \otimes V \to V$ are graded symmetric, e.g. for $n = 2$,

$$\hat{l}_2(A, B) = (-1)^{\deg(A)\deg(B)}\hat{l}_2(B, A).$$

The definition of $\hat{f}_4$ requires more care since it involves an integration over the part of the moduli space covered by ν_4 described above in (3.11). We will come back to this in more generality in Sect. 3.3. For now, let us just state that the world-sheet conformal field theory provides a map from the set of geometric vertices $\{\nu_k, k \geq 3\}$ to the set of multilinear maps $\{\hat{l}_k, k \geq 2\}$. In addition, $\partial \mapsto Q$ will be explained in Sect. 3.3. The correct generalization of the free, closed string action (3.9) is then given by

$$S[\psi] = \frac{1}{2}\Omega(\psi, Q\psi) + \frac{1}{3!}\Omega(\psi, \hat{l}_2(\psi, \psi)) + \frac{1}{4!}\Omega(\psi, \hat{l}_3(\psi, \psi, \psi)) + \cdots .$$

There are two apparent obstacles in interpreting this action as a BV action. First, Ω has ghost number -5, while in the BV formalism one usually assumes a degree -1 odd symplectic form. Second, the physical string fields have ghost number 2, whereas in the BV formalism the physical fields are taken to have degree 0. Both of these issues can be resolved by defining the degree to be the ghost number minus two, so that the classical closed string field is a degree zero element in[1] $V[-2] :=\downarrow^2 V$. To simplify the notation, we will denote $V[-2]$ by V again when there is no risk of confusion. The odd symplectic structure of closed string field theory $\omega : V \otimes V \to \mathbb{C}$ is then identified as

$$\omega := \Omega \circ (\uparrow^2 \otimes \uparrow^2) = \langle -, c_0^- - \rangle \circ (\uparrow^2 \otimes \uparrow^2) .$$

Due to the shift and the c_0^--insertion, ω is graded anti-symmetric and has degree -1. Similarly, the maps l_n are redefined as $l_n =\downarrow^2 \circ \hat{l}_n \circ \uparrow^{\otimes 2n}$. With this convention the maps l_n all have degree 1.

As mentioned above, an important difference against the point particle involves the gauge invariance which was trivial for the latter. At the linearized level,

$$\delta_\Lambda \psi = Q\Lambda$$

with Λ an arbitrary element in V of degree -1, will leave $S[\psi]$ invariant since $Q^2 = 0$. Clearly, this gauge invariance is reducible due to $\Lambda \to \Lambda + Q\Lambda'$, where Λ' has degree -2, etc. This leads to ghosts for ghosts, etc. Luckily this complication

[1]See Appendix A.1 and Part II for the definition of $\uparrow$ and $\downarrow$.

will not affect us here, so we only refer the reader to the original literature listed at the end of this chapter for further details on this issue. To the quadratic order, we have the invariance under

$$\delta_\Lambda \psi = Q\Lambda + l_2(\psi, \Lambda) + \cdots ,$$

because $\partial v_3 = 0$ implies

$$[Q, l_2](a, b) = Ql_2(a, b) + l_2(Qa, b) + (-1)^{\deg(a)} l_2(a, Qb) = 0 . \tag{3.12}$$

So far this looks as a Lie algebra type symmetry. However, since l_3 does not commute with Q as is evident from (3.11) due to the correspondence between the geometric vertex v_4 and the operation l_3, and the boundary operator δ and the BRST operator Q, respectively, $\delta_\Lambda \psi$ will receive corrections. Indeed, we will see in Sect. 3.5 that (3.11) implies

$$[Q, l_3] + \frac{1}{2}[l_2, l_2] = 0 , \tag{3.13}$$

where

$$\begin{aligned}[Q, l_3](a, b, c) =& Ql_3(a, b, c) + l_2(Qa, b, c) + (-1)^{\deg(a)} l_3(a, Qb, c) \\ &+ (-1)^{\deg(a)+\deg(b)} l_3(a, b, Qc)\end{aligned}$$

and

$$\begin{aligned}[l_2, l_2](a, b, c) = & l_2(l_2(a, b), c) + (-1)^{\deg(c)(\deg(a)+\deg(b))} l_2(l_2(c, a), b) \\ &+ (-1)^{\deg(a)(\deg(b)+\deg(c))} l_2(l_2(b, c), a) .\end{aligned}$$

Thus (3.13) says that l_2 satisfies the Jacobi identity only up to a Q-exact term, i.e. we are dealing with a Lie algebra up to homotopy, or a homotopy Lie algebra. The gauge invariance at cubic order is

$$\delta_\Lambda \psi = Q\Lambda + l_2(\psi, \Lambda) + l_3(\psi, \psi, \Lambda) \tag{3.14}$$

and the invariant BV action up to quartic terms is given by

$$S[\Psi] = \frac{1}{2}\omega(\Psi, Q\Psi) + \frac{1}{3!}\omega(\Psi, l_2(\Psi, \Psi)) + \frac{1}{4!}\omega(\Psi, l_3(\Psi, \Psi, \Psi)) ,$$

where Ψ, restricted to degree 0, reduces to ψ but contains, in addition, fields of different degrees. The gauge parameter Λ lives in the degree -1 component. The gauge transformation (3.14) is generated by

$$\delta_\Lambda \psi = \{S, \psi\} ,$$

so that the consistency condition for S is the algebraic BV equation

$$\{S, S\} = 0\,.$$

This is the stringy generalization of (2.16).

Before we return to the geometric BV equation, we would like to touch upon the quantum corrections to the classical BV equation (3.11). The qualitatively new feature here is the presence of loops, obtained by connecting two punctures of the same vertex by a propagator. A generic Riemann surface appearing in this way will not be a fundamental vertex since it is obtained through composition of vertices with propagators. In the BV quantization of field theory such loops generically produce what is usually called ultra-violet (UV) divergences arising when the length of such a propagator shrinks to zero. These divergences have to be regularized, for instance, by introducing a cut-off. If the theory in question is renormalizable, there is a suitable redefinition of the vertices such that the cut-off can be removed. In string theory the situation is different. Indeed, due to the modular invariance of the world-sheet conformal field theory, the part of the moduli space corresponding to a very short, collapsed handle, is equivalent, by a modular transformation, to a region where this loop propagator is long. This feature is at the origin of the UV-finiteness of string theory. One way of implementing this feature in the decomposition of $\hat{\mathscr{P}}_{g,n}$ is to demand that any Jordan curve on the punctured Riemann surface, not homotopic to a point, has length bigger or equal 2π. The length is measured with respect to the metric of minimal area for a Riemann surface of genus g with n-punctures. This data together with the coordinate curves homotopic to the punctures define the moduli space $\hat{\mathscr{P}}_{g,n}$. Note that, according to this prescription, the punctures are replaced by stubs of finite length.

One possible obstruction to this decomposition of the moduli space is that connecting stubs by a propagator of a non-negative length does not cover (or overcovers) the moduli space at genus $g+1$ and $n-2$ punctures. In this case, inserting a propagator of zero length introduces an artificial boundary which needs to be compensated by introducing an elementary vertex of genus $g+1$ and $n-2$ punctures that covers the missing region in the moduli space such that its boundary cancels the one introduced by the zero-length propagator. Thus, we have to extend the geometric BV equation (3.11), introducing an operation Δ that corresponds to the twist-sewing of two punctures of the same vertex with a zero-length propagator. We will write the corresponding BV equation as

$$\begin{aligned}
&\partial \nu_{g+1,1} + \hbar \Delta \nu_{g,3} = 0 \\
&\partial \nu_{g+1,2} + \hbar \Delta \nu_{g,4} = 0 \\
&\partial \nu_{g+1,3} + \hbar \Delta \nu_{g,5} + \frac{1}{2} \sum_{g_1+g_2=g+1} \{\nu_{g_1,2}, \nu_{g_2,2}\} = 0, \quad \text{etc.,}
\end{aligned}$$

where g denotes the genus of the geometric vertex and we introduced the expansion parameter $\hbar$ to keep track of the loop order. It turns out that the operations ∂, Δ, and $\{-,-\}$ satisfy the axioms of a BV algebra, namely

$$\begin{aligned}
&\partial^2 = 0\\
&\Delta^2 = 0\\
&\partial\Delta + \Delta\partial = 0\,,\\
&\partial\{\,-,\;-\} - \{\partial-,\;-\} + \{\,-,\partial-\} = 0\\
&\Delta\{-,-\} - \{\Delta,-\} + \{-,\Delta\} = 0\\
&\{\nu,\mu\} + (-1)^{(\deg(\nu)+1)(\deg(\mu)+1)}\{\mu,\nu\} = 0\\
&(-1)^{(\deg(\nu)+1)(\deg(\rho)+1)}\{\{\nu,\mu\},\rho\} + \text{cycl.} = 0.
\end{aligned}$$

Here ν, μ, and ρ are geometric vertices and $\deg(\nu)$ will be defined in the next section. For instance, $\Delta^2 = 0$ follows from the fact that the sewing increases dimensionality by one due to the twist angle, and that the chains are endowed with an orientation.

3.3 Decomposition of the Moduli Space

Let us now describe the decomposition of the moduli space $\hat{\mathscr{P}}_{g,n}$ more thoroughly. The geometric vertices $\nu_{g,n}$ with labeled punctures are elements of a proper chain complex $C^\bullet(\hat{\mathscr{P}}_{g,n})$ endowed with an orientation.[2] We will not go into details here. The grading is defined by the co-dimension (therefore, we use upper index for the chain degree)

$$\deg(\nu_{g,n}) = \dim(\mathscr{M}_{g,n}) - \dim(\nu_{g,n})\,,$$

where $\mathscr{M}_{g,n}$ is the moduli space of punctured Riemann surfaces of genus g with n punctures; its dimension is $6g + 2n - 6$. With this grading, the boundary operator ∂ has degree one. Furthermore, the twist-sewing defined in the last section induces the operations

$$_a\bullet_b\;: C^{k_1}(\hat{\mathscr{P}}_{g_1,n_1+1}) \times C^{k_2}(\hat{\mathscr{P}}_{g_2,n_2+1}) \to C^{k_1+k_2+1}(\hat{\mathscr{P}}_{g_1+g_2,n_1+n_2}) \tag{3.15}$$

and

$$\bullet_{ab}\;: C^{k}(\hat{\mathscr{P}}_{g,n+2}) \to C^{k+1}(\hat{\mathscr{P}}_{g+1,n}) \tag{3.16}$$

[2]See references in the section on Further Reading for details.

of degree 1 on the complex of singular chains. Here a and b denote the punctures that are being twist sewed.

We can implement the indistinguishability of identical particles already at the geometric level by requiring the invariance under permutations of punctures. The complex of chains that are invariant under permutations of punctures is denoted by $C^\bullet_{\text{inv}}(\hat{\mathscr{P}}_{g,n})$. The maps ${}_a\bullet_b$ and $\bullet_{ab}$ can be lifted to operations on $C^\bullet_{\text{inv}}(\hat{\mathscr{P}}_{g,n})$. This gives rise to

$$\{\nu_{g_1,n_1+1}, \nu_{g_2,n_2+1}\} = \sum_{\sigma \in \text{uSh}(n_1,n_2)} \sigma(\nu_{g_1,n_1+1}\, {}_a\bullet_b\, \nu_{g_2,n_2+1}),$$

where $\text{uSh}(n_1, n_2)$ is the set of (n_1, n_2)-unshuffles, i.e. permutations $\sigma \in \Sigma_{n_1+n_2}$ such that

$$\sigma(1) < \cdots < \sigma(n_1) \;\text{ and }\; \sigma(n_1+1) < \cdots < \sigma(n_1+n_2),$$

and

$$\Delta\nu_{g,n} = \bullet_{ab}\nu_{g,n+2}\,.$$

It is possible to define a graded commutative, associative product of degree 0 on the chain complex of moduli spaces of disconnected surfaces by the disjoint union $\nu \sqcup \mu$ of vertices. Alltogether,

$$(\nu \sqcup \mu) \sqcup \rho = \nu \sqcup (\mu \sqcup \rho)\,, \quad \nu \sqcup \mu = (-1)^{|\nu||\mu|}\mu \sqcup \nu\,, \quad |\nu \sqcup \mu| = |\nu| + |\mu|\,.$$

With respect to this multiplication, Δ is a nilpotent second order derivation, i.e.

$$\Delta^2 = 0,$$

$$\Delta(\nu \sqcup \mu \sqcup \rho) - \Delta(\nu \sqcup \mu) \sqcup \rho - (-1)^{|\nu|}\nu \sqcup \Delta(\mu \sqcup \rho) - (-1)^{(|\nu|+1)|\mu|}\mu \sqcup \Delta(\nu \sqcup \rho)$$
$$+ \Delta(\nu) \sqcup \mu \sqcup \rho + (-1)^{|\nu|}\nu \sqcup \Delta(\mu) \sqcup \rho + (-1)^{|\nu|+|\mu|}\nu \sqcup \mu \sqcup \Delta(\rho) = 0\,.$$

We recognize the structure of a BV algebra. The bracket $\{-,-\}$ expressed in terms of Δ through

$$\{\nu, \mu\} := (-1)^{|\nu|}\Delta(\nu \sqcup \mu) - (-1)^{|\nu|}\Delta(\nu) \sqcup \mu - \nu \sqcup \Delta(\mu)\,,$$

defines a Gerstenhaber bracket, also called an anti-bracket, with the properties

$$\{\nu, \mu\} + (-1)^{(|\nu|+1)(|\mu|+1)}\{\mu, \nu\} = 0\,,$$
$$(-1)^{(|\nu|+1)(|\rho|+1)}\{\{\nu, \mu\}, \rho\} + \text{cycl} = 0\,.$$

Hence, our BV algebra has an underlying Gerstenhaber algebra equipped with the bracket $\{-,-\}$ and multiplication $\sqcup$.

The additional structure needed to ensure that the decomposition of the moduli space $\hat{\mathscr{P}}_{g,n}$ does not produce an artificial boundary (or multiple covering) is a boundary operator ∂, together with the BV master equation

$$\partial \nu_{g,n} + \frac{1}{2} \sum_{n_1 \leq n_2; g_1 \leq g_2} \{\nu_{g_1,n_1+1}, \nu_{g_2,n_2+1}\} + \hbar \Delta \nu_{g-1,n+2} = 0 \tag{3.17}$$

where $n = n_1 + n_2$ and $g = g_1 + g_2$. This geometric decomposition of the moduli space is manifestly background independent, in particular, independent of the choice of a metric in M.

Remark 3.2 In string theory, the background dependence enters through the conformal field theory morphism that maps this structure to the BV algebra of the physical Hilbert space. As explained above, the Polyakov action on a given space-time defines a conformal field theory. Its BRST quantization introduces the Faddeev–Popov ghosts c and b. The resulting BRST symmetry is generated by a ghost number one BRST differential Q. In contrast to the point particle discussed in the first chapter, in string theory, the construction of Q depends crucially on the choice of a background. For a generic choice of the metric g on M the path integral measure in (3.3) fails to be conformally invariant so that the construction presented here cannot be used.

3.4 Measure, Vertices, and BV Action

In order to complete the construction of the CFT morphism between the geometric and algebraic BV structures, we need to construct a measure on $\hat{\mathscr{P}}_{g,n}$. One way to do this is by completing the set of BRST transformations (3.5) by

$$\delta_{BRST} h_{ab} = \delta h_{ab} , \quad \delta_{BRST} \delta h_{ab} = 0 ,$$

where h_{ab} is the reference metric and δh_{ab} is a ghost number 1, traceless, symmetric tensor. This extended set of BRST transformations is a symmetry of the evolution kernel (3.3) provided we add the term

$$\Delta I = \frac{1}{4\pi i} \int_{\Sigma} \sqrt{h}\, \delta h_{ab} b^{ab}\, d\tau d\sigma$$

to the action in (3.3). To continue, we consider the generating functional

$$F_{\mathscr{A}}(h, \delta h) = \int D[\phi, b, c]\, e^{-I-\Delta I} \mathscr{A}(\phi, c), \tag{3.18}$$

where $\mathscr{A}(\phi, c)$ is a polynomial in vertex operators obtained, according to the operator-state correspondence, by gluing unit disks with the corresponding vertex operators at their origin. These states may be off-shell and therefore vertex operators are not assumed to be primaries. Now, if we let n_c and n_b be the number of c-ghost zero modes and the number of b-ghost zero modes, respectively, we have, because of the ghost number anomaly, that $n_b - n_c = 6g - 6$. Consequently, upon expanding (3.18) in δh we find that $F_{\mathscr{A}}(h, \delta h)$ is a homogeneous function of degree at least $6g - 6$ in δh which, due to the anticommuting nature of δh, will be seen to give rise to a differential form on the space of complex structures $\mathscr{J}$ of a Riemann surface $\Sigma_{g,n}$ of genus g with n punctures. Furthermore, the exterior derivative of $F_{\mathscr{A}}$ is just the BRST differential

$$dF_{\mathscr{A}} = \int_{\Sigma} \sum_{a,b} \delta h_{ab} \frac{\delta F_{\mathscr{A}}}{\delta h_{ab}} = Q\, F_{\mathscr{A}} \,. \tag{3.19}$$

It is possible to show that $F_{\mathscr{A}}(h, \delta h)$ is the pullback of a differential form on the moduli space $\hat{\mathscr{P}}_{g,n}$ of punctured Riemann surfaces $\Sigma_{g,n}$ together with a choice of a coordinate curve around each puncture. Let $\mathscr{D}_{p_1,\cdots,p_n}$ denote the group of orientation-preserving diffeomorphisms that are trivial to all orders near the points $p_1, \cdots, p_n$. Then $\hat{\mathscr{P}}_{g,n} = \mathscr{J}/\mathscr{D}_{p_1,\cdots,p_n}$ parametrizes not only the punctures but the coordinate curves homotopic to these punctures as well. By construction $F_{\mathscr{A}}(h, \delta h)$ is invariant under $\mathscr{D}_{p_1,\cdots,p_n}$. Furthermore, if v is an infinitesimal vector field representing one of the generators of $\mathscr{D}_{p_1,\cdots,p_n}$, then its contraction with $F_{\mathscr{A}}(h, \delta h)$ vanishes. Indeed, because

$$\delta h_{ab} \to \delta h_{ab} + \epsilon(v_{a;b} + v_{b;a}) \,,$$

the contraction of $F_{\mathscr{A}}(h, \delta h)$ with v is given by

$$\delta F_{\mathscr{A}}(h, \delta h) = \int D[\phi, b, c]\, e^{-I-\Delta I}\, \mathscr{A}(\phi, c) \int_{\Sigma} \epsilon(v_{a;b} + v_{b;a}) b^{ab} \,,$$

which is equivalent to a shift in the c-ghost, $\delta c = \epsilon v$. Since $\mathscr{A}(\phi, c)$ depends only on $c(p_i)$, $i = 1, \cdots, n$, and their derivatives at the punctures where v vanishes to all orders, it follows from the translation invariance of the ghost measure that the contraction of $F_{\mathscr{A}}(h, \delta h)$ with v vanishes, too. This proves that $F_{\mathscr{A}}(h, \delta h)$ is the pullback of a form on $\hat{\mathscr{P}}_{g,n}$. In order to integrate $F_{\mathscr{A}}(h, \delta h)$, one would like to choose a suitable parameterization $\{m_s\}$, $s = 1, \cdots \dim(\nu_{g,n})$, of the subspace $\nu_{g,n}$ of $\hat{\mathscr{P}}_{g,n}$. This change of variables will induce a Jacobian through

$$\delta h_{ab} = \sum_{s=1}^{\dim(\nu_{g,n})} \frac{\partial h_{ab}}{\partial m_s} dm_s \,.$$

Using the tracelessness of b^{ab}, we can implement this as a product (over s) of

$$J_s = \frac{1}{4\pi i}\int_\Sigma \frac{\partial(\sqrt{h}\,\delta h_{ab})}{\partial m_s} b^{ab}\, d^2 z \tag{3.20}$$

into the path integral measure in (3.3). Using the Schiffer variation argument, we can represent a tangent vector on $\nu_{g,n}$ by a collection of n Witt vectors. To illustrate this procedure, we consider again the tetrahedron in Fig. 3.5. The idea is to cut out a disc around one of the four punctures, deforming it by the flow generated by the Witt vector and finally to sew it back in. In the case at hand, we know that there exists one non-vanishing meromorphic vector field w defined in a neighborhood of that puncture, which cannot be extended to the whole sphere. These vector fields generate translations in the moduli space, i.e. they move the punctures and deform the coordinate curves around them. Upon substitution into (3.20) and using the equation of motion $\nabla_i b^{ij} = 0$, we find

$$J = \frac{1}{2\pi i}\oint b(w)\,.$$

More generally, in case of p punctures the Jacobian is $J_p = \prod_{s=1}^{p} J_s$, so we end up with the p-form on $\nu_{g,n}$

$$F(m) = \mathscr{N}\left(\int D[\phi, b, c]\, e^{-I}\, \mathscr{A}(\phi, c)\, J_p\right) dm_1 \wedge \cdots \wedge dm_p\,,$$

where $\mathscr{N}$ denotes a normalization. A top form is obtained for

$$p = \dim(\nu_{g,n}) = 6g - 6 + 2n,$$

where n is the number of punctures on Σ with vertex operators of ghost number 2 inserted at each puncture. In addition, due to (3.19) and the BRST invariance of the path integral measure in (3.3) under $I \to I + \Delta I$, we obtain an important identity

$$dF_{\mathscr{A}}(m) + F_{Q\mathscr{A}}(m) = 0\,, \tag{3.21}$$

which expresses the fact that $F_{\mathscr{A}}(m)$ defines a chain map. An immediate consequence of this is that the multilinear maps

$$\hat{f}_n^g = \int_{\nu_{g,n}} F_{\mathscr{A}} \in \mathrm{Hom}_{inv}(V^{\otimes n}, \mathbb{C}) \tag{3.22}$$

satisfy an algebraic BV master equation. Indeed, for $\mathscr{A} \in V^{\otimes n}$ we have

$$\begin{aligned} Q\hat{f}_n^g &= \int_{\mathcal{V}_{g,n}} F_{Q\mathscr{A}} = \int_{\mathcal{V}_{g,n}} dF_{\mathscr{A}} = \int_{\partial^{\text{geo}}\mathcal{V}_{g,n}} F_{\mathscr{A}} \\ &= -\frac{1}{2} \sum_{n_1 \le n_2; g_1 \le g_2} \int_{\{\mathcal{V}_{g_1,n_1+1}, \mathcal{V}_{g_2,n_2+1}\}^{\text{geo}}} F_{\mathscr{A}} - \hbar \int_{\Delta^{\text{geo}}\mathcal{V}_{g,n+2}} F_{\mathscr{A}} \\ &= -\frac{1}{2} \sum_{n_1 \le n_2; g_1 \le g_2} \{\hat{f}_{n_1+1}^{g_1}, \hat{f}_{n_2+1}^{g_2}\}^{\text{alg}} - \hbar \Delta^{\text{alg}} \hat{f}_{n+2}^g , \end{aligned} \tag{3.23}$$

where we used in the second line the geometric BV master equation (3.17). The operation

$$\{-, -\}^{\text{alg}} : \text{Hom}_{inv}(V^{\otimes n_1}, \mathbb{C}) \times \text{Hom}_{inv}(V^{\otimes n_2}, \mathbb{C}) \to \text{Hom}_{inv}(V^{\otimes n_1+n_2-2}, \mathbb{C}) \tag{3.24}$$

was defined in Chap. 2 via the contraction of inputs of $\hat{f}_{n_1}^{g_1}$ and $\hat{f}_{n_2}^{g_2}$ with the inverse of the symplectic form ω. Similarly,

$$\Delta^{\text{alg}} : \text{Hom}_{inv}(V^{\otimes n}, \mathbb{C}) \to \text{Hom}_{inv}(V^{\otimes n-2}, \mathbb{C}) \tag{3.25}$$

is, for $n_1, n_2 \ge 1$, and $n \ge 2$, defined through the contraction of two inputs of $\hat{f}_n^g$ with ω^{-1}. The subscript inv is to emphasize that $\hat{f}_n^g$ is graded symmetric under the permutation of its inputs. In what follows, we will drop the label "alg" unless there is a danger of confusion with the geometric operations introduced before.

We are now ready to write down the complete quantum BV action of closed string field theory. Let $f_n^g = \hat{f}_n^g \circ (\uparrow^2)^{\otimes n}$ denote the algebraic vertex of genus g with n closed string insertions. This vertex comes with a certain power in $\hbar$, namely $2g + n/2 - 1$. The full BV action then reads

$$S(\Psi) = \sum_{g,n \ge 2} \hbar^{2g+n/2-1} f_n^g(\Psi) ,$$

where Ψ is the closed string field. Furthermore, $f_2^0(\Psi)$ can be identified with the quadratic action (3.9).

Remark 3.3 It is worth pointing out that, while in the decomposition (3.17) the boundary operator ∂ has a well-defined action on the vertices, it does not act on the punctures. This distinction does not appear on the image of the chain map (3.21) where the action of Q on the vertices is induced by its action on the vector space attached to the puncture.

As a consequence of (3.23), the action S satisfies the quantum BV master equation

$$\hbar \Delta S + \frac{1}{2}\{S, S\} = 0\,, \tag{3.26}$$

where the action of Δ and $\{-, -\}$ was defined in (3.24) and (3.25), respectively. This equation will be recovered, with different notation, in (8.36) of Part II.

3.5 Algebraic Structure

We want to interpret the vertices of S in the language of homotopy algebras. Let $TV = \bigoplus_{n=0}^{\infty} V^{\otimes n}$ be the tensor algebra and $\mathrm{Hom}^{cycl}(TV, V)$ the space of graded symmetric maps from TV to V. Since ω is non-degenerate and the vertices are invariant with respect to any permutation of the inputs, there is a unique map l_n^g : $V^{\otimes n} \to V$ such that

$$f_n^g(\Psi) = \frac{1}{n!}\omega(l_{n-1}^g(\Psi^{\otimes n-1}), \Psi)\,, \quad g \geq 0\,,$$

where $l_n^g = l^g \circ i_n$, with i_n the inclusion map $V^{\otimes n} \to TV$. The map l^g is an element of $\mathrm{Hom}^{cycl}(TV, V)$, i.e.

$$\omega(a_1, l_n^g(a_2, \ldots, a_{n+1}) = (-1)^{|a_1|+|a_2|+|a_1|(|a_2+\cdots+a_{n+1}|)}\omega(a_2, l_n^g(a_3, \ldots, a_{n+1}, a_1)\,.$$

Upon substitution into the closed string BV bracket we get

$$\begin{aligned}\{S, S\} &= \frac{\partial_l S}{\partial \Psi^i}\omega^{ij}\frac{\partial_r S}{\partial \Psi^j} \\ &= \sum_{\substack{g_1, g_2 \geq 0 \\ n_1, n_2 \geq 1}} \frac{\hbar^{2g_1+2g_2-1}}{n_1! n_2!}\omega\left(e_i, l_{n_1}^{g_1}(c^{\otimes n_1})\right)\,\omega^{ij}\,\left(e_j, l_{n_2}^{g_2}(c^{\otimes n_2})\right)\,,\end{aligned}$$

where $c \equiv \hbar^{1/2}\Psi$, $\Psi = \Psi^i e_i$ for $\{e_i\}$ a homogeneous basis of V. Here, we took into account the sign $(-1)^{-i}$, with i the degree of e_i, when commuting ∂_{Ψ^i} through l_n^g, which, in turn, is compensated by

$$\omega(\Psi, l_n^g(\Psi, \cdots, \Psi, e_i, \Psi, \cdots, \Psi)) = (-1)^i\omega(e_i, l_n^g(\Psi, \cdots, \Psi)),$$

due to the cyclicity of the functions l_n^g. With $\omega_{ij} = \omega(e_i, e_j)$ and $\delta^i{}_j = \omega^{ik}\omega(e_k, e_j)$, we obtain

$$\{S, S\} = \sum_{\substack{g_1,g_2\geq 0\\ n_1,n_2\geq 1}} \frac{\hbar^{2g_1+2g_2-1}}{n_1!n_2!}\omega\left(l_{n_2}^{g_2}(c^{\otimes n_2}), l_{n_1}^{g_1}(c^{\otimes n_1})\right)$$

$$= \sum_{\substack{g_1,g_2\geq 0\\ n_1,n_2\geq 1}} \frac{\hbar^{2g_1+2g_2-1}}{n_1!n_2!}\omega\left(c, l_{n_2}^{g_2}(l_{n_1}^{g_1}(c^{\otimes n_1}), c^{\otimes n_2-1})\right)$$

where the second sum runs over all $(n_1, n_2 - 1)$-unshuffles σ and where we denote by $\epsilon(\sigma)$ the Koszul sign, cf. (1) of Part II. In the present case, this latter sum merely results in the factor $\frac{1}{n!}$ since all inputs are identical and of total degree zero. Similarly,

$$\hbar\Delta S = \sum_g \frac{\hbar^{2g-1}}{(n-2)!}\omega(c, l_{n-1}^{g-1}(e^i \otimes e_i \otimes c^{\otimes n-3}))\,,$$

where $e^i = \omega^{ij}e_j$. Altogether, the BV equation (3.26) is equivalent to

$$\sum_{\substack{g_1+g_2=g\\ n_1+n_2=n-1}} \sum_\sigma \epsilon(\sigma) l_{n_2}^{g_2}(l_{n_1}^{g_1}(c_{\sigma(1)}, \ldots, c_{\sigma(n_1)}), c_{\sigma(n_1+1)}, \ldots, c_{\sigma(n-1)})$$
$$+ l_{n+1}^{g-1}(e^i \otimes e_i \otimes c^{\otimes n-1}) = 0\,, \tag{3.27}$$

which, in turn, is recognized as the loop homotopy algebra described is detail in Sect. 8.2, see (8.42) in particular.

An equivalent shorthand expression of the condition (3.27) in terms of the graded symmetrized inputs is

$$\sum_{\substack{g_1+g_2=g\\ i_1+i_2=n}} l_{i_1+1}^{g_1} \circ (l_{i_2}^{g_2} \wedge \mathbb{1}^{\wedge i_1})(e^c) + \hbar\, l_{n+2}^{g-1}(\omega^{-1} \wedge \mathbb{1}^{\wedge n})(e^c) = 0\,, \tag{3.28}$$

where $e^c := \sum\limits_{n=0}^{\infty} \frac{1}{n!}c^{\wedge n}$.

3.6 Coalgebra Description

In the scaling limit $\hbar \to 0$ with $c \equiv \hbar^{1/2}\Psi$ finite, (3.28) reproduces the defining equations of a strongly homotopy Lie or L_∞-algebra, described in more detail in Appendix A and in Part II. A convenient way to define L_∞-algebras is the cobar construction where one collects the various maps into a single object

$$L = \sum_{n=1}^{\infty} \sum_{i+j=n} \sum_{\sigma}{}' (l_i \wedge \mathbb{1}^{\wedge j}) \circ \sigma \,, \tag{3.29}$$

where $\sum_\sigma'$ indicates the sum over all unshuffles $\sigma \in \Sigma_n$ with

$$\sigma_1 < \cdots < \sigma_i, \quad \sigma_{i+1} < \cdots < \sigma_n.$$

The permutation σ denotes the map that sends $c_1 \wedge \cdots \wedge c_n$ to $(-1)^\epsilon c_{\sigma_1} \wedge \cdots \wedge c_{\sigma_n}$. As reviewed in Appendix A, L is a coderivation of $SV = \bigoplus_{n=0}^{\infty} V^{\wedge n}$. The space Coder$(SV)$ is equipped with the canonical bracket $[-,-]$ defined through the composition of maps, cf. Appendix A. For $\hbar = 0$, (3.28) is equivalent to

$$L^2 = \frac{1}{2}[L, L] = 0,$$

which is precisely the defining relation for an L_∞-algebra. We recognize a theory with vertices given by a cyclic L_∞-algebra L as a classical closed string theory.

Remark 3.4 The construction of the minimal model map for classical closed string theory proceeds in close analogy with the point particle. We already noticed in Sect. 2.2 that the construction of the minimal model is equivalent to the construction of tree-level S-matrix amplitudes via Feynman rules. First one chooses a gauge (e.g., the Siegel gauge $b_0^+\psi = 0$) so that one can define a propagator Q^{-1}. With the aid of the latter, we construct all possible rooted trees with vertices labeled by $l_n := \pi_1 \circ L \circ i_n$ and internal edges labeled by the propagator. Here π_1 stands for the projection to maps with one output. The collection of all these trees, with inputs and the output restricted to the cohomology H, then defines the multilinear maps $\tilde{l}_n = \pi_1 \circ \tilde{L} \circ i_n$ in complete analogy with Sect. 2.2. Thus, $\tilde{l}_n$ represents the $n+1$-string S-matrix amplitudes. Furthermore, $\tilde{L}^2 = 0$ so that the map

$$F \,: (H, \tilde{L}) \to (V, L)$$

is a quasi-isomorphism between L_∞-algebras. The proof of these claims goes analogically with the point particle case in Sect. 2.2, with

$$c = F(c^P) = \pi_1(\mathbb{1} - l_1^{-1}\delta)^{-1}(e^{\wedge c^P} - \mathbb{1})\,,$$

where $c^P \in H$ and

$$\tilde{l}_n = P\,\delta(\mathbb{1} - l_1^{-1}\delta)^{-1} i_n\,, \quad n \geq 2\,, \tag{3.30}$$

where $\delta = \pi_1 \circ L \circ (\mathbb{1} - i_1)$. Then

$$(Q + \delta)F = F\tilde{L} \qquad \text{and} \qquad \tilde{L}^2 = 0$$

can be established by inspection. More generally, this structure follows from the homological perturbation lemma. On physical grounds, the latter property is a consequence of the Ward identities implied by the gauge invariance.

Let us now return to the quantum theory, $\hbar \neq 0$, whose algebraic structure is that of the loop homotopy algebra (also called quantum L_∞-algebra). It can be described in terms homomorphisms $l_i^g : V^{\wedge i} \to V$, with an additional label given by the genus g, as well as the inverse ω^{-1} of the symplectic form ω as in (3.28). Alternatively, one can consider ω^{-1} as one of the operations of the algebra. In this setup we have not only operations with one output like l_i^g, but also the operation ω^{-1} with no input and two outputs. From this point of view, this algebraic structure is a special case of a homotopy involutive Lie bialgebra, also called an *IBL*$_\infty$-algebra. We give a detailed account of such algebras in Appendix A and in Part II. To describe their relation to the quantum BV equation, we lift ω^{-1} to a coderivation $D(\omega^{-1}) \in \text{Coder}^2(SA)$ of order two defined by (see Appendix A)

$$\pi_1 \circ D(\omega^{-1}) = 0 \quad \text{and} \quad \pi_2 \circ D(\omega^{-1}) = \omega^{-1}\,. \tag{3.31}$$

The combination

$$\mathfrak{L} = \sum_{g=0}^{\infty} \hbar^g \sum_{n=1}^{\infty} \sum_{i+j=n} \sum_{\sigma}{}' (l_i^g \wedge \mathbb{1}^{\wedge j}) \circ \sigma \; + \hbar D(\omega^{-1}) \tag{3.32}$$

defines an element in $\text{Coder}(SA, \hbar)$ of degree 1. Condition (3.28) is then equivalent to

$$\mathfrak{L}^2 = 0\,,$$

where the cyclicity of $\mathfrak{L}$ with respect to ω_c is assumed. These are the algebraic relations of quantum closed string field theory expressed in terms of *IBL*$_\infty$-algebras.

3.7 Uniqueness and Background Independence

In this section, we will discuss the question whether the CFT morphism that determines the algebraic string field theory vertices and, therefore, the particular realization of an L_∞-algebra, is unique up to an equivalence. This important question in string theory is, however, not directly relevant to the rest of this book and can thus be skipped by the reader who is more interested in the algebraic aspects of string theory.

There is a definite answer to the above question as far as continuous deformations of a given SFT are concerned. Let us denote the classical closed string vertices by $l_n \equiv l_n^0$. The bracket $[-,-]$ on Coder(SA) (see Appendix A) induces the Chevalley–Eilenberg differential $d_C = [L,-]$ on the deformation complex. On the other hand, due to the isomorphism $\mathrm{Hom}(V_c^{\wedge k}, V) \to \mathrm{Hom}(V_c^{\wedge k+1}, \mathbb{C})$ this induces a differential d_c on the cyclic complex. Thus, any consistent infinitesimal deformation $\Delta l = \{\Delta l_n\}_{n\in\mathbb{N}}$ of the L_∞-structure $\{l_n\}_{n\in\mathbb{N}}$ (Δ not to be confused with the BV operator) is d_c-closed, $d_c(\Delta l) = 0$. By carefully analyzing continuous deformations of the CFT morphism, one arrives at the following result:

Let $I[\phi, c, \bar{c}, b, \bar{b}]$ be the closed string world-sheet CFT action on M defining the morphism from the BV algebra of the geometric vertices $\{v_n\}$ to the BV algebra of algebraic vertices s_n, V_c the corresponding module of conformal tensors (closed string Hilbert space) and Q_c the BRST differential. Then

$$\mathrm{coh}(d_c) = \emptyset\,.$$

Remark 3.5 The above result has important consequences for the background independence of classical closed string field theory. It is intimately related to the nature of equivalence classes of L_∞-algebras. Clearly, L_∞-field redefinitions preserve the L_∞-structure and can be interpreted as $[\cdot,\cdot]$-gauge symmetry transformations if they are continuously connected to the identity. On the other hand, field redefinitions include shifts in the closed string background. These are easily seen to be L_∞-isomorphisms. For a given homotopy algebra we can then consider a non-vanishing Maurer–Cartan element Ψ_0 with $L(e^{\Psi_0}) = 0$, in order to construct a twisted homotopy algebra $L_{\Psi_0} = E(-\Psi_0) \circ L \circ E(\Psi_0)$ upon conjugation. The definition of $E(-\Psi_0)$ is given in Appendix A. The background independence then would imply that the structure maps of the minimal model obtained from this homotopy algebra are equivalent to the perturbative S-matrix elements of the world-sheet CFT in the new background, see Fig. 3.7.

Since $coh(d_c) = \emptyset$, the L_∞-algebra $L_{CFT(\Psi_0)}$ obtained from the world-sheet theory in the new background Ψ_0 corresponding to the MC-element Ψ_0 is L_∞-equivalent to the L_∞-algebra L_{Ψ_0} obtained from L by conjugation, i.e.

$$L_{CFT(\Psi_0)} = K^{-1} \circ L_{\Psi_0} \circ K\,,$$

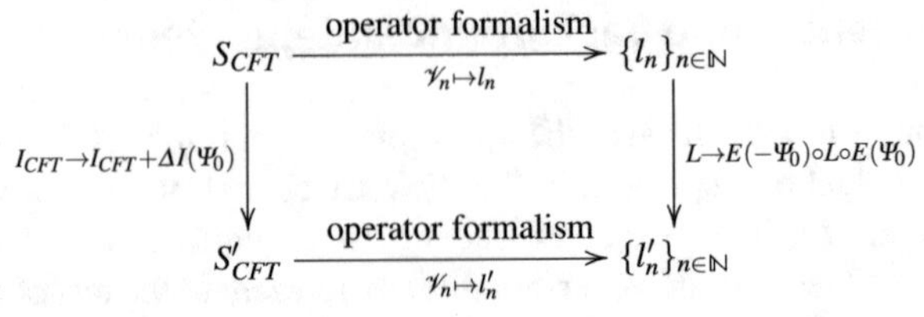

Fig. 3.7 A schematic representation of the background independence. The vertical arrow on the left represents a deformation of the background implemented by a deformation $\Delta I(\Psi_0)$ of the CFT morphism

where K is an L_∞-isomorphism continuously connected to the identity. However, since the L_∞ equivalence classes identify all continuously connected closed backgrounds, we cannot conclude from the above that $L_{CFT(\psi_0)}$ and L_{ψ_0} actually describe the same background. The necessary refinement for this is then provided by the open-closed homotopy algebra described in the next section, which implies that $K = \mathbf{1}$. We should note, however, that generic on-shell closed string backgrounds may not be continuously connected to each other and furthermore do not preserve V_c. This puts a limitation on applicability of the proof of background independence given here.

3.8 Summary, Comments, and Remarks Towards Part II

It should be clear from the discussion in this chapter, that abstractly, quantum closed string field theory can be defined as a morphism from the BV algebra of singular chains on the moduli space of punctured closed Riemann surfaces to the BV algebra of functions on a differential graded vector space V, endowed with a degree -1 symplectic form ω, i.e. on the state space of the first quantized closed string. Such a morphism realized by a world-sheet conformal field theory carries geometric vertices $\nu_{g,n}$ assembled into a solution of the quantum BV master equation (3.17) on the space of chains on the moduli space into algebraic vertices $f_{g,n}$ defining the string field theory action. Geometric vertices themselves are determined by a decomposition of the moduli space.

The relation to Part II is as follows. The decomposition of the moduli space can be interpreted as a morphism between two odd (aka twisted) modular operads, see Definition 6.24. This morphism goes from the Feynman transform, cf. Definition 7.9, of the modular commutative operad in Definition 6.18 into the odd modular operad of chains on the moduli space. On the latter there is an obvious action of permutations of punctures. The operadic operations (3.15) and (3.16) are induced by the twisted sewing and self-sewing of punctured closed Riemann surfaces.

We know from Theorem 8.2 that such a morphism is equivalent to a solution of the quantum BV equation on the respective space of invariants under permutations. In the case of the commutative operad, this is the complex of chains invariant under the permutations of punctures as described in Sect. 3.3. This is the horizontal arrow in Fig. 1.1 providing the geometric vertices $\nu_{g,n}$.

The BV algebra morphism given by formula (3.22) that sends a geometric vertex $\nu_{g,n}$ into the algebraic vertex f_n^g, when extended from invariant cochains to all cochains, can again be interpreted as a morphism of two odd modular operads going from the odd modular operad of cochains on the moduli space to the endomorphism operad, i.e. as a representation of the former one. This is the vertical arrow in Fig. 1.1 providing the algebraic vertices f_n^g.

Finally, the composition of the two odd modular operad morphisms gives the algebra over the Feynman transform of the modular commutative operad, i.e. the diagonal arrow in our Fig. 1.1. As such, it is equivalent to a solution to the quantum BV master equation on the space of complex-valued graded symmetric maps on the CFT state space, described by algebraic vertices $f_n^g : V^{\otimes n} \to \mathbb{C}$, due to Theorem 8.3 of Part II. This is, of course, guaranteed by the construction that uses the composition of the horizontal and vertical arrows in Fig. 1.1. The result is the quantum closed SFT action in Zwiebach's construction. Using the odd symplectic form, the algebraic vertices f_n^g can be turned into degree 1 n-ary brackets $l_{g,n}$: $V^{\otimes n} \to V$. The resulting homotopy algebra is the loop homotopy algebra (aka quantum L_∞-algebra) described in Sect. 3.5 of Part I and again from the operadic point of view in Sect. 8.2 of Part II. The graded symmetry of the vertices/brackets is due to the trivial action of permutations on the commutative operad.

The same algebraic structure, i.e. the loop homotopy algebra can be reinterpreted in terms of an IBL_∞-algebra. The main difference is that the BV operator now, contrary to the loop homotopy algebra formulation, becomes one of the operations. In this formalism the quantum BV master equation $\hbar\Delta + \frac{1}{2}\{S, S\} = 0$ equivalently translates to the nilpotency of the full BV operator $\hbar\Delta + \{S, -\}$. IBL_∞-algebras, as they are used here, are described in Appendix to Part I, cf. Sect. B.2. A more rigorous alternate version is given in Part II, Sect. 8.3.

All constructions described above have their classical analogue. In STF it is the scaling limit $\hbar \to 0$, $c \equiv \hbar^{1/2}\Psi$, cf. Sect. 3.6, leading to the classical BV action and the corresponding L_∞-algebra. Since the powers of $\hbar$ count loop contributions, we may forget about loops.

In the operadic language of Part II, this scaling limit corresponds to taking the forgetful functor from the modular commutative operad (the right adjoint to the modular envelope functor (6.18), cf. Sect. 6.4), to the cyclic commutative operad. Roughly speaking, as above, we forget about loops everywhere and consider only genus 0 corollas. The horizontal arrow in Fig. 1.1 now goes from the cobar construction of the latter (here we consider only the trees generated by corollas instead of all graphs) into the odd cyclic operad of the genus 0 closed Riemannian surfaces with punctures, where the self-sewing operation is now forbidden. The vertical arrow is as before given by conformal field theory, however, now evaluated only at zero genus. The composition, i.e. the diagonal arrow in Fig. 1.1, is the classical (genus 0) closed string field theory.

Of course, one may wonder how Fig. 1.1 modifies in the world of IBL_∞-algebras. This would lead us to properads, their cobar construction and representations. We will, however, not pursue this here.

Concerning the scattering amplitudes, i.e. the induced loop homotopy structure on the Q-cohomology, this works very much the same way as already described in Chap. 1, cf. (2.35). Although, in formula (3.30), we gave only the result for (cyclic) L_∞-algebras, the modification of that formula to loop homotopy algebras is straightforward. Note that (3.30) is the formula from the homological perturbation lemma. It expresses the change of the trivial differential on functions on the Q-cohomology H induced by the perturbation of the BRST operator Q by the map δ introduced there. Now, the loop homotopy version leads to a similar formula, e.g. L in δ is replaced by $\hbar\Delta+L$, where we think of the BV operator Δ as an operation with zero inputs and two outputs and where L comprises also all higher genus operations. As in Remark 2.10, it is enough for our purposes to think about morphisms of loop homotopy algebras as of (nonlinear) maps between the underlying vector spaces compatible with the symplectic structures and intertwining between the respective full BV operators.

Further Reading

There are many good textbooks for the world-sheet formulation of string theory which is relevant, in particular, for the calculation of scattering amplitudes. Standard textbooks are

- M. B. Green, J. H. Schwarz and E. Witten, "Superstring theory. Vol. 1 and 2", Cambridge Monographs On Mathematical Physics, (1987),
- J. Polchinski, "String theory. Vol. 1 and 2", Cambridge University Press (2005),
- R. Blumenhagen, D. Luest and S. Theisen, "Basic concepts of string Theory", Springer Verlag (2013).

For early work on BRST invariant string field theory, see

- W. Siegel and B. Zwiebach, "Gauge string fields", Nucl.Phys. B263 (1986) 105–128, and
- C. B. Thorn, "String field theory", Phys. Rept. 175, 1–101 (1989).

For the covariant formulation of interacting quantum closed string field theory, see

- B. Zwiebach, "Closed string field theory: Quantum action and the B-V master equation", Nucl.Phys. B390, (1993) 33–152, and
- B. Zwiebach, "Oriented open - closed string theory revisited", Annals Phys. 267, (1998) 193–248.

Recently, an alternative decomposition of the closed string moduli space in terms of hyperbolic string vertices has been proposed in

- K. Costello and B. Zwiebach, "Hyperbolic string vertices", arXiv:1909.00033.

The first proof of local background invariance of closed string field theory appeared in

- A. Sen and B. Zwiebach, "A proof of local background independence of classical closed string field theory", Phys. B414 (1994) 649–714.

Our description on this topic follows

- K. Munster and I. Sachs, "Homotopy classification of bosonic string field theory", Commun. Math. Phys. 330 (2014) 1227–1262.

In our description of the measure on the moduli space we followed Witten's notes

- E. Witten, "Perturbative superstring theory", arXiv:1209.5461.

For a modern review of closed superstring field theory, see, e.g.

- C. de Lacroix H. Erbin, S. P. Kashyap, A. Sen, M. Vermaet, "Closed superstring field theory and its applications", Int. J. Mod. Phys. A32, (2017), and
- T. Erler, "Four lectures on closed string field theory", arXiv:1905.06785.

A derivation of the L_∞-structure of closed string field theory can be found in

- H. Kajiura and J. Stasheff, "Homotopy algebras inspired by classical open-closed string field theory", Commun. Math. Phys. 263, 553–581 (2006).

For a detailed description of the IBL_∞-algebra underlying quantum closed string theory see

- K. Cieliebak, K. Fukaya and J. Latschev, "Homological algebra related to surfaces with boundary", arXiv:1508.02741.

One description of the appropriate complex on the moduli space of (bordered) Riemann surfaces with punctures can be found in

- E. Harrelson, A.A. Voronov and J. J. Zuniga, "Open-closed moduli spaces and related algebraic structures", Lett. Math. Phys. 94 (2010), no. 1, 1–26.

Open and Closed Strings

4

4.1 World-Sheets with Boundaries

In addition to the theory of closed strings just described, we may construct another consistent theory by considering world-sheets Σ with boundaries. Such a theory necessarily contains open strings. In this case, we can insert open string states to the boundaries and closed string states to the bulk of Σ as in Fig. 4.1.

Let us first characterize the open string Hilbert space V_o. For this, we need to fix the boundary conditions for the various fields in the world-sheet CFT. Let us focus on Σ of genus 0 with one boundary and no closed string insertions. Hence, the world-sheet of an open string is topologically the infinite strip $[0, \pi] \times \mathbb{R}$. By the conformal mapping $z = -e^{-iw}$ ($w = \sigma + i\tau$ with $(\sigma, \tau) \in [0, \pi] \times \mathbb{R}$), the strip is mapped to the upper half plane $\mathbb{H}$. For concreteness, we take Neumann boundary conditions $\partial_n \phi(z, \bar{z})|_{\partial\Sigma} = 0$, and similarly for c or b in place of ϕ. Here, ∂_n is the derivative in the direction normal to the boundary. Alternate (e.g. Dirichlet) boundary conditions will not affect our treatment below as long as they preserve the conformal symmetry of the world-sheet action, that is, admissible boundary conditions must satisfy the Cardy condition $(T - \bar{T})|_{\partial\Sigma} = 0$. The fields living on $\mathbb{H}$ can be separated into holomorphic and anti-holomorphic parts, but due to the boundary conditions, these two parts combine into a single holomorphic field defined on the whole complex plane $\mathbb{C}$. This is the doubling trick; the holomorphic field in the lower half plane is expressed in terms of the anti-holomorphic field in the upper half plane through the reflection of the coordinate, $\partial_z \phi(z)|_{\mathrm{Im}(z<0)} = \partial_{\bar{z}} \phi(\bar{z})|_{\mathrm{Im}(z>0)}$, etc. We then expand each field on $\mathbb{C}$ in a Laurent series (mode expansion)

$$i\partial_z \phi(z) = \sum_{n \in \mathbb{Z}} \frac{\alpha_n}{z^{n+1}}, \quad c(z) = \sum_{n \in \mathbb{Z}} \frac{c_n}{z^{n-1}}, \quad b(z) = \sum_{n \in \mathbb{Z}} \frac{b_n}{z^{n+2}},$$

© Springer Nature Switzerland AG 2020

M. Doubek et al., *Algebraic Structure of String Field Theory*, Lecture Notes in Physics 973, https://doi.org/10.1007/978-3-030-53056-3_4

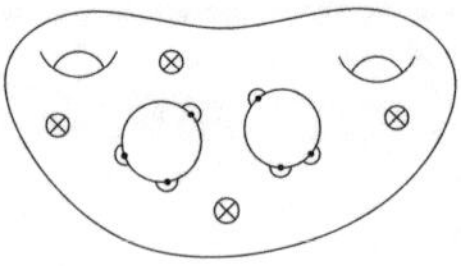

Fig. 4.1 Open-closed world-sheet with open string punctures on the boundary and closed string punctures in the bulk

where the conformal weights are $h_{\partial_z\phi} = 1$, $h_c = -1$, $h_b = 2$, and the modes satisfy the commutation relations

$$[\alpha^\mu_m, \alpha^\nu_n] = m g^{\mu\nu}\delta_{m+n,0}\,, \quad \{c_m, b_n\} = \delta_{m+n,0}\,.$$

Similarly,

$$L_n = \oint \frac{dz}{2\pi} z^{n+1} T(z)\,, \qquad T(z)|_{\mathrm{Im}(z<0)} = \bar{T}(\bar{z})|_{\mathrm{Im}(z>0)}\,.$$

In the operator formalism, the space of states V_o is generated by acting with the creation operators on the vacuum $|0, \mathbf{k}\rangle$. The grading on V_o is again induced by assigning ghost number one to c, minus one to b and zero to ϕ, i.e. every c mode increases the ghost number by one, whereas the b modes decrease the ghost number by one. The operator-state correspondence for open string states works much the same as for the closed case by mapping $\tau = -\infty$ to $z = 0$. We then define the corresponding Fock space as the set of polynomials in the fields and their derivatives $\Phi(\phi, b, c, \partial_z\phi, \cdots)$ at the origin. The vector $\Phi[\bar{\phi}, \bar{b}, \bar{c}] \in V_o$ is then defined by evaluating the path integral measure in (3.3) on the half disk with $\Phi(\phi, b, c, \partial_z\phi, \cdots)$ inserted at the origin and subject to boundary conditions at $|z| = 1$

$$\Phi[\bar{\phi}, \bar{b}, \bar{c}] = \int\limits_{\{\phi,b,c\}|_{\partial\mathbb{H}}=\{\bar{\phi},\bar{b},\bar{c}\}} D[\phi, b, c]\, e^{-I(\phi.b.c.h)}\,\Phi(\phi, \partial_z\phi, \cdots)(0)\,.$$

A physical open string state is then a ghost number one state that has a representative as a conformal primary with conformal weight $h = 0$ of the form

$$\Phi(0) = \Phi(\phi(0), \partial_z\phi(0), \cdots)c(0)\,,$$

where the $c(0)$-insertion represents the globally defined vector field on $\mathbb{H}$ that is non-vanishing at $z = 0$ and which generates the translation of the origin. Nonphysical states correspond to non-primary insertions with arbitrary dependence on c and its derivatives. Utilizing the operator-state correspondence, we can identify every state $\phi \in \tilde{V}_o$ with a local operator $\mathcal{O}_\phi$ and define the BPZ inner product by

$$\langle \Phi_1, \Phi_2 \rangle := \lim_{z\to 0} \big\langle (I^*\mathcal{O}_{\Phi_1})(z)\mathcal{O}_{\Phi_2}(z)\big\rangle_H\,,$$

where $I(z) = -1/z$, $\langle \dots \rangle_H$ is the correlator on the upper half plane and $I^*\mathcal{O}$ denotes the conformal transformation of $\mathcal{O}$ with respect to I. Due to the $PSL(2, \mathbb{R})$-invariance this correlator is non-vanishing only if it is saturated by three c-ghost insertions and consequently, the BPZ inner product carries ghost number -3. The BRST invariance of the open string is the same as for the closed string modulo the identification of the left and right moving modes as explain above. Thus,

$$Q = \oint \frac{dz}{2\pi i}\, c(z) \left(T^{\phi}(z) + \frac{1}{2} T^{g}(z) \right) = c_0 L_0 + \cdots .$$

In this way we find the kinetic term of the open string field action

$$S_{kin} = \tfrac{1}{2}\langle \Phi, Q_o \Phi \rangle$$

which, upon variation, reproduces the correct cohomology for the open string physical states. This is best seen in Siegel gauge, $b_0 \Phi = 0$.

In order to identify the BPZ inner product with the odd symplectic structure ω, we shift the degree by one, which turns an odd graded symmetric inner product into an odd symplectic structure

$$\omega_o := \langle -, - \rangle \circ (\uparrow \otimes \uparrow) : V_o[-1] \otimes V_o[-1] \to \mathbb{C} ,$$

where $V_o[-1] :=\downarrow V_o$. In what follows, we will not distinguish between $V_o[1]$ and V_o.

To summarize, we have an odd symplectic structure ω_o on V_o of degree -1 and the classical open string field which is a degree zero element in V_o.

Having constructed the kinetic term for the open string, let us now turn to the interactions of open and closed strings. Topologically, the generic elementary interaction vertex, sketched in Fig. 4.2, is a genus g Riemann surface Σ with b boundaries, m punctures on the boundaries, and n punctures in the interior. The corresponding moduli space is denoted by $\mathcal{M}_{n,m}^{b,g}$. A geometric vertex in $\mathcal{M}_{n,m}^{b,g}$ of real dimension $6g - 6 + 2n + 3b + m$ is defined by the metric of minimal area under the condition that the length of any nontrivial open curve in Σ with endpoints at the boundaries be greater or equal to π and that the length of any nontrivial closed curve be greater or equal to 2π. The geometric decomposition of the moduli space of bordered Riemann surfaces is again described by a master equation of the from (3.17) although the details of the sewing procedure are more involved.

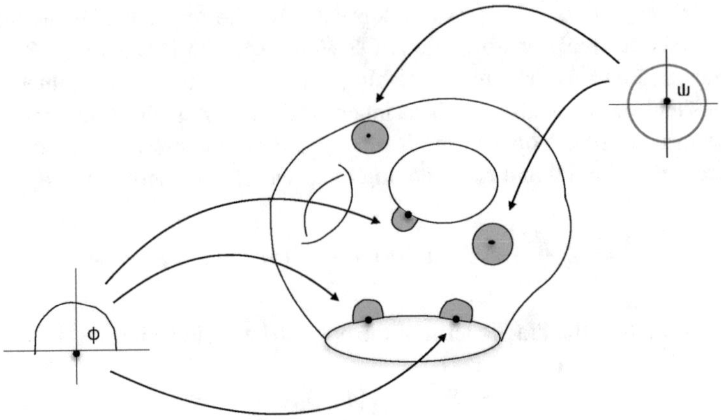

Fig. 4.2 Open-closed vertex with closed and open states inserted by sewing in unit disks and semidisks along the respective coordinate curves

4.2 Open String

Let us first consider world-sheets with a single boundary of genus zero and no closed string punctures. The geometric cubic interaction vertex is a disk with three punctures and local coordinates around the punctures. This vertex has no modulus in analogy with the cubic closed string vertex. We can build a Feynman diagram for the scattering of four strings by inserting a propagator between any two punctures of two cubic vertices. This certainly covers the part of the moduli space of the disk with four punctures. In fact, it turns out that these Feynman diagrams cover the whole moduli space $\mathscr{M}_{4,0}^{1,0}$. In analogy to the closed string, we would now like to express this fact as a solution of a BV equation. Since the geometric vertex is invariant under a rotation of the punctures on the boundary, it seems natural to assume that the cyclic permutations act trivially on the cochains of $\mathscr{M}$. However, it turns out that it is not possible to define a BV bracket $\{-, -\}$ on this moduli space. We therefore consider the cyclic cochain complex instead for which

$$\nu(p_2, \cdots, p_n, p_1) = (-1)^{n-1}\nu(p_1, \cdots, p_{n-1}, p_n)\,. \tag{4.1}$$

Later, we will see that this choice is the correct one also from the point of view of the CFT-morphism to the actual open string theory vertices. On the cyclic complex, the bracket is defined by sewing the last puncture p_{n_1} of ν with the first one of μ

$$\{\nu, \mu\} = (-1)^{m_\nu d_\mu}(\nu\;{}_{p_{n_1}}\!\bullet_{q_1}\mu)_{\text{cycl}}\,, \tag{4.2}$$

where m_ν is the number of punctures of ν, d_μ is the dimension of the moduli space associated with μ, and the subscript "cycl" stands for the sum over all cyclic permutations ensuring that the result is again in the cyclic complex. As an

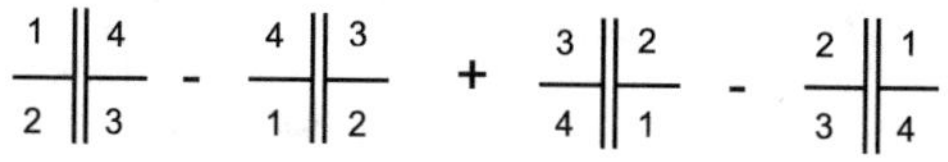

Fig. 4.3 The BV bracket of two cubic open string vertices

example, let us compute $\{\nu_3, \nu_3\}$. First we note that ν_3 is just a point. Thus $d_{\nu_3} = 0$. We then glue the third puncture of the first cubic vertex (straight line) with the first puncture of the second vertex (straight line) and subsequently sum over the cyclic permutations of the four punctures with the signs coming from (4.1). This is sketched in Fig. 4.3 and is easily recognized as twice the difference between the s-channel and t-channel amplitude for the scattering $(1, 2) \to (3, 4)$. From this we also see that the graphs constructed by connecting cubic vertices by propagators cover the moduli space for the disk with four punctures. Indeed, since the t-channel smoothly crosses into the s-channel when the propagators collapse to zero length, there is no boundary at that point and therefore no ν_4 is needed. Thus, the geometric BV equation that underlies the classical ($\hbar = 0$) field theory of open strings is simply

$$\{\nu_3, \nu_3\} = 0 \,. \tag{4.3}$$

For the closed string, where cyclic permutations are replaced by the action of the full symmetric group, there would be an additional u-channel contribution, altogether 4! terms.

We should stress that the action of the operator Δ, defined by the operation of gluing the first puncture with the second, plus gluing the first puncture with the third, etc., followed by a subsequent cyclic summation, increases the number of boundaries and therefore takes us outside the restriction assumed here. So, the field theory of open strings described here is a classical one ($\hbar = 0$). Apart from this restriction, this defines a consistent string field theory. The CFT morphism is constructed in complete analogy with Sect. 3.4 and induces the action

$$S = \tfrac{1}{2}\langle \Phi, Q_o \Phi\rangle + \frac{1}{3}\langle \Phi, \Phi * \Phi\rangle \,, \tag{4.4}$$

where $\Phi * \Phi$, given by the image of ν_3 via the CFT morphism, defines an associative product due to (4.3). This is just the open string field theory of Witten. Its algebraic structure is simply that of a differential graded associative algebra $(V_o, *, Q_o)$ together with an invariant inner product, $\langle - - \rangle$ on V_o, that is the one that satisfies

$$\langle a * b, c\rangle = \langle a, b * c\rangle \;\text{ and }\; \langle a, b\rangle = (-1)^{|a||b|}\langle b, a\rangle .$$

Remark 4.1 Classically, consistent deformations of this open string field theory are given by homotopy associative, or A_∞-algebras $(V_o, \hat{m}_k)$, $k \geq 1$, that preserve the inner product. To see this, we use the fact that, in the presence of an invariant inner product, there is a natural isomorphism $\mathrm{Hom}(A^{\otimes k}, A) \to \mathrm{Hom}(A^{\otimes k+1}, \mathbb{C})$. This can be used to write the action as

$$S_o = \sum_{n=1}^{\infty} \frac{1}{n+1} \omega_o(m_n(\Phi^{\otimes n}), \Phi) .$$

Here, $\Phi = \Phi^i e_i$, where $\{e_i\}$ is a homogeneous basis of V_o. To keep the notation concise, we denote the degree $|e_i|$ of e_i simply by i and degree of Φ^i by $-i$. The classical BV equation applied to an open string field theory action then gives

$$\begin{aligned} \{S_o, S_o\}_o &= \frac{\partial_l S_o}{\partial \Phi^i} \omega^{ij} \frac{\partial_r S_o}{\partial \Phi^j} \\ &= \sum_{n_1=1}^{\infty} \sum_{n_2=1}^{\infty} \omega_o \left(e_i, m_{n_1}(\Phi^{\otimes n_1})\right) \omega^{ij}\, \omega_o \left(e_j, m_{n_2}(\Phi^{\otimes n_2})\right) , \end{aligned} \tag{4.5}$$

where we used that ∂_{Φ^i} picks up a sign $(-1)^i$ when commuted through m_n as well as the equality

$$\omega_o(\Phi, m_n(\Phi, \cdots, \Phi, e_i, \Phi, \cdots, \Phi)) = (-1)^i \omega_o(e_i, m_n(\Phi, \cdots, \Phi)).$$

Then, using $\omega_{ij} = \omega(e_i, e_j)$ and $\delta^i{}_j = \omega^{ik}\omega(e_k, e_j)$, we find

$$\begin{aligned} \{S_o, S_o\}_o &= \sum_{n_1=1}^{\infty} \sum_{n_2=1}^{\infty} \omega_o \left(m_{n_2}(\Phi^{\otimes n_2}), m_{n_1}(\Phi^{\otimes n_1})\right) \\ &= \sum_{n=1}^{\infty} \frac{2}{n+1} \sum_{i+j+k=n} \omega_o \left(m_{i+k+1}\left(\Phi^{\otimes i} \otimes m_j(\Phi^{\otimes j}) \otimes \Phi^{\otimes k}\right), \Phi\right) \\ &= \sum_{n=1}^{\infty} \frac{2}{n+1} \omega_o \left(\pi_1 \circ M^2(\Phi^{\otimes n}), \Phi\right) = 0 . \end{aligned}$$

All we had to use was the cyclicity of m_n. Also, $M \in \mathrm{Coder}^{cycl}(TA_o)$ is the coderivation corresponding to $m \in \mathrm{Hom}^{cylc}(TA_o, A_o)$ and Eq. (4.5) is equivalent to $M^2 = 0$, the well-known statement that the vertices of a classical open string field theory define an A_∞-algebra. We give a detailed description of A_∞-algebras in Appendix A. Their definition is very similar to that of L_∞-algebras described in the previous section.

Infinitesimal deformations of this type of algebras are governed by cyclic cohomology. To see this connection, we note that in the presence of an invariant inner product, there is a natural isomorphism $\mathrm{Hom}(A^{\otimes k}, A) \to \mathrm{Hom}(A^{\otimes k+1}, \mathbb{C})$. The image of $\hat{m}_k \in \mathrm{Hom}(V_o^{\otimes k}, V_o)$ in $\mathrm{Hom}(V_o^{\otimes k+1}, \mathbb{C})$ is

$$f_{k+1}(\Phi_1, \cdots, \Phi_{k+1}) = \omega(\hat{m}_k(\Phi_1, \cdots, \Phi_k), \Phi_{k+1}) .$$

The cyclic symmetry (4.1) implies that $\hat{m}_k$ preserves the inner product. For example,

$$\begin{aligned}\omega(\hat{m}_2(\Phi_1, \Phi_2), \Phi_3) &= f_3(\Phi_1, \Phi_2, \Phi_3) = (-1)^{|\Phi_1|(|\Phi_2|+|\Phi_3|)} f_3(\Phi_2, \Phi_3, \Phi_1) \\ &= (-1)^{|\Phi_1|(|\Phi_2|+|\Phi_3|)} \omega(\hat{m}_2(\Phi_2, \Phi_3), \Phi_1) \\ &= \omega(\Phi_1, \hat{m}_2(\Phi_2, \Phi_3)) ,\end{aligned}$$

where we used (4.1) in the second identity and $|\Phi|$ denoted the ghost number of the string field Φ. In close analogy to the closed string, the collection of maps $\{\hat{m}_k\}$ can be lifted to a coderivation on

$$T V_o = \bigoplus_{n=0}^{\infty} V_o[1]^{\otimes n} .$$

Then the bracket $[M, -]$, $M = \sum m_n$, on $\mathrm{Coder}(T V_o)$ induces a Hochschild coboundary operator d_H on $CC^k \equiv \mathrm{Hom}_{\mathrm{cycl}}(V_o^k, V_o)$. For the cubic theory at hand, $\hat{m}_1 = Q$ and $\hat{m}_2 = *$, we have

$$d_H = (-1)^{k+1} Q + \delta ,$$

where

$$(Q f_k)(\Phi_1, \ldots, \Phi_k) = \sum_{i=1}^{k} (-1)^{\Phi_1 + \ldots + \Phi_{i-1}} f_k(\Phi_1, \ldots, Q\Phi_i, \ldots, \Phi_k)$$

and

$$\begin{aligned}(\delta f_k)(\Phi_1, \cdots, \Phi_{k+1}) = &\sum_{i=1}^{k} (-1)^i f_k(\Phi_1, \cdots, \Phi_i * \Phi_{i+1}, \cdots, \Phi_{k+1}) \\ &+ (-1)^{\Phi_1(\Phi_2 + \cdots + \Phi_{k+1})} f_k(\Phi_2, \cdots, \Phi_k, \Phi_{k+1} * \Phi_1) .\end{aligned}$$

From the above, it is clear that d_H takes cyclic elements to cyclic elements. Furthermore, d_H squares to zero since $Q^2 = \delta^2 = [Q, \delta] = 0$. The cyclic cohomology is then the cohomology of d_H in CC^*. It turns out that this cohomology is isomorphic to the physical cohomology of closed strings. More precisely:

Let $I[\phi, c, b]$ be the world-sheet action defining open string world-sheet CFT-morphism between geometric and algebraic vertices, V_o the corresponding module *of conformal tensors (open string Hilbert space) and Q the open string BRST differential. Then, the only nontrivial infinitesimal deformations of this CFT-morphism preserving V_o are infinitesimal deformations of the closed string background in the relative cohomology of Q_c*

$$coh(d_H) \cong coh(b^-, Q_c) \,. \tag{4.6}$$

This can alternatively be expressed by saying that the only way to deform open string theory is to place it in a nontrivial closed string background, e.g. a curved metric on $\mathbb{R}^{26}$ that solves the closed string equations of motion.

Remark 4.2 A particular class of deformations that do not preserve Q_o are the shifts in the open string background. Such transformations are, however, d_H-exact as are all field redefinitions of ϕ. From a physics perspective, the interesting fact implied by the above result is that open string theory already contains the complete information about the particle content of *closed* string theory.

4.3 Summary, Comments, and Remarks Towards Part II

Here, we comment only on open strings and leave the comments on the open-closed theory to Sect. 5.3. Most of the discussion below will be parallel to the case of the closed string in Sect. 3.8. Concerning physics, the main difference is that there is no consistent open string filed theory beyond the classical level, i.e. the one containing only interaction through a disc with (at least three) punctures on the boundary. A consistent quantum field theory of open strings needs the inclusion of closed strings. In the previous sections, we described the classically consistent Witten's open SFT (4.4), which contains only cubic interaction, and where the resulting algebraic structure is that of a differential graded cyclic associative algebra. Classically consistent deformations of this open string theory lead us to a generalization of associative structures, i.e. to their homotopy versions, which are the cyclic A_∞-algebras. Obviously, the vertices are graded symmetric only with respect to cyclic permutations of punctures on the disc boundary.

In the operadic language of Part II, A_∞-algebras are algebras over the cobar construction of the cyclic associative operad of Example 6.18. Here, the difference against the cyclic commutative operad is the absence of a nontrivial action of all permutations. The consequence is the above-mentioned graded symmetry of the corresponding algebraic vertices f_k and of the corresponding products m_k, in Sect. 4.2, under the cyclic permutations only.

Concerning the construction of Witten's open SFT in the spirit of Fig. 1.1, in the upper left corner we would have the cobar construction of the cyclic associative operad. In the upper right corner we would have the odd cyclic operad of cyclic chains (4.1) on the moduli space of discs with punctured boundaries, the operadic operations being defined by sewing discs via punctures as in (4.2). The horizontal arrow, a morphism of these two operads, would be given by a decomposition of the moduli space. In the present case, the relevant decomposition uses only the trivalent geometric vertex ν_3 satisfying the classical BV master equation (4.3). The horizontal arrow, provided by the open genus-0 CFT and landing in the endomorphism operad of the open conformal theory state space, gives the cubic open STF action (4.4).

Despite the non-existence of a consistent quantum open string field theory, nothing prevents us from constructing an associative analogue of a loop homotopy algebra (quantum L_∞-algebra). We may call it a quantum A_∞-algebra, or quantum open homotopy algebra, similarly as we could have coined the name quantum closed homotopy algebra for a quantum L_∞-algebra. Actually, quantum versions of homotopy algebras will exist for all modular operads and also for their colored versions, e.g. quantum open-closed homotopy algebra over the two-colored combination of the modular associative and modular commutative operad.

Hence, in the quantum associative case, we start from the modular associative operad, i.e. the modular envelope of the cyclic associative operad. It is the linearization of the modular operad described by Theorem 6.1 in Part II. Axioms of quantum homotopy associative algebras are then obtained by applying Theorem 8.2 to its Feynman transform.

Also, for quantum A_∞-algebras, in particular for their classical versions, i.e. A_∞-algebras, we could, using homological perturbation lemma, transfer the respective algebra structures to their Q-cohomology, cf. the corresponding discussion in Sect. 2.3 related to formulas (2.35) and (3.30), respectively.

Finally, let us make a comment on D-branes, a subject which we did not touch at all. These can be included in the classical A_∞-picture. We refer the interested reader to work of Gaberdiel and Zwiebach cited below.

Further Reading

The open bosonic string field theory described in this chapter was formulated by E. Witten in

• E. Witten, "Noncommutative geometry and string field theory", Nucl. Phys. B268 (1986) 253

The proof that the cubic action (4.4) does realize a decomposition of the moduli space of open Riemann surfaces with punctures on the boundary can be found in

• B. Zwiebach, "A proof that Witten's open string theory gives a single cover of moduli space", Commun. Math. Phys. 142 (1991) 193.

A derivation of the isomorphism (4.6) can be found in

• N. Moeller and I. Sachs, "Closed String Cohomology in Open String Field Theory", JHEP 1107 (2011) 022.

The algebraic structure of open string theory with non-abelian gauge groups was discussed in

- M. R. Gaberdiel and B. Zwiebach, "Tensor Constructions of Open String Theories I: Foundations", Nucl.Phys. B505 (1997) 569–624, and
- M. R. Gaberdiel and B. Zwiebach, "Tensor Constructions of Open String Theories II: Vector bundles, D-branes and orientifold groups", Phys.Lett. B410 (1997) 151–159.

5 Open-Closed BV Equation

An important conclusion at the end of the previous chapter is that the unique consistent infinitesimal deformation of classical open string field theory is an open-closed vertex with one closed string puncture, cf. the italicized paragraph before formula (4.6). To continue, we want to analyze the consistency of a generic open-closed vertex as in Fig. 4.1. For this, we first need to review the various sewing operations on Riemann surfaces with labeled boundaries and labeled punctures in the bulk as well as on the boundaries. Concerning the sewing of closed string punctures, we have already discussed it in Sect. 3.3. The sewing of two open string punctures on different vertices with the corresponding cyclic complex was treated in the previous section. What remains, is the sewing of open string punctures on the same surface. Acting with the geometric operator Δ on two open punctures on the same boundary of a given Riemann surface Σ increases the number of boundaries by one and decreases the number of open string punctures by two while leaving the genus invariant. In contrast, acting on two punctures on different boundaries of the same surface increases the genus by one, decreases the number of boundaries by one and decreases the number of punctures by two. A detailed discussion of the various possible sewings of open-closed surfaces can be found in the literature quoted at the end of this chapter. Through the present chapter, we use notation and terminology introduced in the appendix to Part I.

5.1 Open-Closed BV Action

The operations described above can again be packaged into a geometric BV equation with a degree one bracket $\{-, -\}$ and a degree one BV operator provided we consider the singular chain complex with a factor $(-1)^{(m_i+1)(m_j+1)}$ assigned under the exchange of boundary i with boundary j of the same surface. Here m_i

© Springer Nature Switzerland AG 2020
M. Doubek et al., *Algebraic Structure of String Field Theory*, Lecture Notes in Physics 973, https://doi.org/10.1007/978-3-030-53056-3_5

and m_j are the numbers of open string punctures on the respective boundaries. The BV equation then reads

$$\partial \nu^{b,g}_{n,m} + \frac{1}{2} \sum_{\substack{n_1 \le n_2; g_1 \le g_2 \\ b_1 \le b_2; m_1 \le m_2}} \{\nu^{b_1,g_1}_{n_1+1,m_1}, \nu^{b_2,g_2}_{n_2+1,m_2}\}_c + \hbar \Delta_c \nu^{b,g-1}_{n+2,m}$$

$$+ \frac{1}{2} \sum_{\substack{n_1 \le n_2; g_1 \le g_2 \\ b_1 \le b_2; m_1 \le m_2}} \{\nu^{b_1,g_1}_{n_1,m_1+1}, \nu^{b_2,g_2}_{n_2,m_2+2}\}_o + \hbar \Delta_o \nu^{b-1,g}_{n,m+2} + \hbar \Delta_o \nu^{b+1,g-1}_{n,m+2} = 0\,,$$

where $n = n_1 + n_2$, $g = g_1 + g_2$, $b = b_1 + b_2$, and $m = m_1 + m_2$. Here and in what follows, we will decorate with the subscript c objects that are associated with closed strings (V_c, Δ_c, etc.) and with the subscript o objects that are associated with open strings (V_o, Δ_o, etc.),

A solution to this BV equation can be obtained using the minimal area construction of Zwiebach: Given a genus g Riemann surface Σ with b boundaries, m punctures on the boundaries and n punctures in the interior, the string vertex is defined by the metric of minimal area under the condition that the length of any non-trivial open curve in Σ with endpoints at the boundaries be greater or equal to π and that the length of any non-trivial closed curve be greater or equal to 2π.

In order to transfer this BV-structure from the complex of geometric vertices to the BV-algebra on $V_o \oplus V_c$, one needs an extension of the path integral measure defining $F_{\mathscr{A}}$ in (3.18) to include surfaces with boundaries. This is straightforward and we will thus not enter in the details. The resulting string field theory vertex $f^{b,g}_{n,m_1,\dots,m_b}$ of genus g with n closed string insertions and b boundaries with m_i representing the number of insertions on the i-th boundary comes with the power $2g + b + n/2 - 1$ in $\hbar$. The full BV action reads

$$S(c,a) = \sum_{b,g} \sum_{n} \sum_{m_1,\dots,m_b} \hbar^{2g+b+n/2-1} f^{b,g}_{n,m_1,\dots,m_b}(c,a)\,, \tag{5.1}$$

where $c \in V_c$ is the closed string field and $a \in V_o$ is the open string field. The BV equation (3.26) puts constraints on the collection of vertices $f^{b,g}_{n,m_1,\dots,m_b}$ and our goal is to interpret these constraints in the language of homotopy algebras.

The idea is to split the set of all vertices into two disjoint sets. One contains all vertices corresponding to closed Riemann surfaces and the other contains the vertices associated with bordered Riemann surfaces. For the former, the action will be given in (5.2). Taking all symmetries of vertices with open and closed inputs into account, we can write the part of the action for the latter as

$$\sum_{n} \sum_{m_1,\dots,m_b} \hbar^{2g+b+n/2-1} f^{b,g}_{n,m_1,\dots,m_b}(c,a) = \frac{1}{b!} \hbar^{2g+b-1} f^{b,g}(e^{\hbar^{1/2}c}; \underbrace{\bar{e}^a, \dots, \bar{e}^a}_{b \text{ times}})\,,$$

where $f^{b,g} \in \text{Hom}(TV_c, \mathbb{C}) \otimes (\text{Hom}^{cycl}(TV_o, \mathbb{C}))^{\wedge b}$. Furthermore, $\bar{e}^a := \sum_{n=1}^{\infty} \frac{1}{n} a^{\otimes n}$ and TV_o denotes the tensor algebra of V_o. To summarize, the full BV-quantum action of open-closed string field theory can be expressed as

$$S = \sum_{g=0}^{\infty} \hbar^{2g-1} \omega_c(l^g, \cdot)(e^{\hbar^{1/2}c}) + \sum_{b=1}^{\infty} \sum_{g=0}^{\infty} \frac{1}{b!} \hbar^{2g+b-1} f^{b,g}(e^{\hbar^{1/2}c}; \underbrace{\bar{e}^a, \ldots, \bar{e}^a}_{b \text{ times}}) .$$

5.2 Quantum Open-Closed Homotopy Algebra

In order to give an algebraic interpretation of the quantum open-closed string field theory, we first have to identify the algebraic structure on V_c and $\text{Hom}^{cycl}(TV_o, \mathbb{C})$. In the last part of this section we will connect the open and closed string part by an IBL_∞-morphism and finally define the quantum open-closed homotopy algebra, that is, the algebraic structure of quantum open-closed string field theory.

As stated in Sect. A.1, the space of cyclic coderivations $\text{Coder}^{cycl}(TA)$ is a Lie algebra, with Lie bracket

$$[D_1, D_2] = D_1 \circ D_2 - (-1)^{|D_1||D_2|} D_2 \circ D_1 .$$

If A is in addition a cyclic A_∞-algebra (A, M, ω), the space $\text{Coder}^{cycl}(TA)$ becomes a dgla, where the differential is defined by $d_H = [M, -]$. First we will transfer the dgla structure from $\text{Coder}^{cycl}(TA)$ to the cyclic Hochschild complex $\mathscr{A} := \text{Hom}^{cycl}(A, \mathbb{C})$. Let $f, g \in \mathscr{A}$, with both having at least one input. We define associated maps in $\text{Hom}^{cycl}(TA, A)$ by

$$\omega(d_f, -) := f , \qquad \omega(d_g, -) := g ,$$

and lift them to cyclic coderivations $D_f, D_g \in \text{Coder}^{cycl}(TA)$. We define the Gerstenhaber bracket on the cyclic Hochschild complex $\mathscr{A}$ by

$$[f, g] := (-1)^{|f|+1} \omega\big(\pi_1 \circ [D_f, D_g], -\big) . \tag{5.2}$$

In the case where one of the maps $f, g \in \mathscr{A}$ has no inputs, we define the commutator to be identically zero. Note that the Gerstenhaber bracket as defined in (5.2) is graded symmetric and has degree one. Thus, the structure induced on $\mathscr{A}$ is a Lie algebra up to a shift in degree, that is, the actual Lie algebra lives on $s\mathscr{A}$. Furthermore, the map that associates a cyclic coderivation to an element of the cyclic Hochschild complex defines a morphism of Lie algebras

$$[D_f, D_g] = (-1)^{|f|+1} D_{[f,g]} .$$

It turns out that we can endow $\mathscr{A}$ with the structure of a differential involutive Lie bialgebra, i.e., that there is a map $\delta : \mathscr{A} \to \mathscr{A}^{\wedge 2}$ such that $d_H, [-,-]$ and δ satisfy the defining Eqs. (B.4)–(B.10) of an *IBL*-algebra. We then define $\delta : \mathscr{A} \to \mathscr{A}^{\wedge 2}$ by

$$
\begin{aligned}
&(\delta f)(a_1, \ldots, a_n)(b_1, \ldots, b_m) \\
&:=(-1)^{|f|} \sum_{i=1}^{n} \sum_{j=1}^{m} (-1)^{\epsilon} f(e_k, a_i, \ldots, a_n, a_1, \ldots, a_{i-1}, e^k, b_j, \ldots, b_m, b_1, \ldots, b_{j-1}) ,
\end{aligned}
$$

where ϵ is the Koszul sign resulting from permutation of the entries, $\{e_k\}$ is a basis of A, and $\{e^k\}$ denotes the corresponding dual basis with respect to the symplectic structure ω. This definition ensures that δf has the right symmetry properties. Furthermore, $d_H, [-,-]$ and δ satisfy all conditions (B.4)–(B.10). Now, let us put this into the language of IBL_∞-algebras. Lift the Hochschild differential, the Gerstenhaber bracket and the cobracket δ to coderivations on $S\mathscr{A}$, the symmetric algebra over $\mathscr{A}$,

$$
\widehat{d_H} \in \mathrm{Coder}(S\mathscr{A}) , \qquad \widehat{[-,-]} \in \mathrm{Coder}(S\mathscr{A}) , \qquad \widehat{\delta} \in \mathrm{Coder}^2(S\mathscr{A}) .
$$

The statement that the maps $d_H, [-,-]$ and δ satisfy the defining relations of a differential *IBL*-algebra is then equivalent to

$$
(\widehat{d_H} + \widehat{[-,-]} + x\widehat{\delta})^2 = 0 ,
$$

where x is a formal expansion parameter which we may identify with $\hbar$. If the algebra A is not endowed with the structure of a cyclic A_∞-algebra, the differential d_H is absent, but we still have an *IBL*-algebra defined by

$$
\mathfrak{L}_o^2 = 0 ,
$$

where

$$
\mathfrak{L}_o := \widehat{[-,-]} + x\widehat{\delta} \in \mathrm{Coder}(S\mathscr{A}, x) \qquad \text{and} \qquad |\mathfrak{L}_o| = 1 . \tag{5.3}
$$

We use Gothic characters for formal power series with values in coderivations. This is the structure that will enter in the definition of the quantum open-closed homotopy algebra. That means that we do not anticipate that the vertices of classical open string field theory define an A_∞-algebra but rather, as we will see soon, derive it from the quantum open-closed homotopy algebra.

Now, we can put the parts together and define the quantum open-closed homotopy algebra (QOCHA). The QOCHA is defined by an IBL_∞-morphisms from the IBL_∞-algebra of closed strings to the *IBL*-algebra of open strings

$$
(V_c, \mathfrak{L}_c) \xrightarrow{IBL_\infty-\text{morphism}} (\mathscr{A}_o, \mathfrak{L}_o) , \tag{5.4}
$$

where $\mathfrak{L}_c \in \mathfrak{coder}(TV_c, \hbar)$ is defined in Eq. (3.32) and $\mathfrak{L}_o \in \text{Coder}(S\mathscr{A}_o, \hbar)$ is defined in Eq. (5.3). We use the abbreviation $\mathscr{A}_o = \text{Hom}^{cycl}(TV_o, \mathbb{C})$. More precisely, we have an IBL_∞-morphism $\mathfrak{F} \in \text{Morph}(V_c, \mathscr{A}_o, \hbar)$, that is,

$$\mathfrak{F} \circ \mathfrak{L}_c = \mathfrak{L}_o \circ \mathfrak{F} \qquad \text{and} \qquad |\mathfrak{F}| = 0. \tag{5.5}$$

The morphism $\mathfrak{F}$ is determined by a map $\mathfrak{f}$ through (see Eq. (B.11) and (B.12))

$$\mathfrak{F} = \sum_{n=0}^{\infty} \frac{1}{n!} \mathfrak{f}^{\wedge n} \circ \Delta_n \ ,$$

where

$$\mathfrak{f} = \sum_{b=1}^{\infty} \sum_{g=0}^{\infty} \hbar^{g+b-1} f^{b,g} \ ,$$

and

$$f^{b,g} : TV_c \to \mathscr{A}_o^{\wedge b} \ .$$

In order to gain a better geometric intuition of (5.5), it is useful to disentangle this equation. First consider the left-hand side of Eq. (5.5). We have

$$\Delta_n \circ L^g = \sum_{i+j=n-1} (\mathbb{1}^{\otimes i} \otimes L^g \otimes \mathbb{1}^{\otimes j}) \circ \Delta_n$$

and

$$\Delta_n \circ D(\omega_c^{-1}) = \sum_{i+j=n-1} \left(\mathbb{1}^{\otimes i} \otimes D(\omega_c^{-1}) \otimes \mathbb{1}^{\otimes j}\right) \circ \Delta_n$$
$$+ \sum_{i+j+k=n-2} \left(\mathbb{1}^{\otimes i} \otimes D(e_i) \otimes \mathbb{1}^{\otimes j} \otimes D(e^i) \otimes \mathbb{1}^{\otimes k}\right) \circ \Delta_n \ ,$$

where D denotes the coderivation of order one defined by

$$\pi_1 \circ D(e_i) = e_i \ .$$

In the following, we abbreviate $L_q = \sum_g \hbar^g L^g$. We get

$$\begin{aligned}\mathfrak{F} \circ \mathfrak{L}_c = & \sum_{n=0}^{\infty} \frac{1}{n!} \sum_{i+j=n-1} (\mathfrak{f}^{\wedge i} \wedge \mathfrak{f} \circ (L_q + \hbar D(\omega_c^{-1})) \wedge \mathfrak{f}^{\wedge j}) \circ \Delta_n \\ & + \sum_{n=0}^{\infty} \frac{1}{n!} \sum_{i+j+k=n-2} \hbar \left(\mathfrak{f}^{\wedge i} \wedge \mathfrak{f} \circ D(e_i) \wedge \mathfrak{f}^{\wedge j} \wedge \mathfrak{f} \circ D(e^i) \wedge \mathfrak{f}^{\wedge k} \right) \circ \Delta_n \\ = & \left(\left(\mathfrak{f} \circ \mathfrak{L}_c + \frac{1}{2} \hbar (\mathfrak{f} \circ D(e_i) \wedge \mathfrak{f} \circ D(e^i)) \circ \Delta \right) \wedge \mathfrak{F} \right) \circ \Delta \, .\end{aligned}$$

Let us turn to the right-hand side of Eq. (5.5). There we have the maps $\widehat{\delta}$ and $\widehat{[-,-]}$. The defining map $\delta = \pi_2 \circ \widehat{\delta}$ of $\widehat{\delta}$ has two outputs and one input. Recall that the order of a coderivation is the number of outputs of the underlying defining map (see Sect. B.1). Similarly, we can define higher order derivations by the number of inputs of the underlying defining map. So we can interpret $\widehat{\delta}$ either as a second order coderivation or as a first order derivation, and $\widehat{[-,-]}$ as a first order coderivation or as a second order derivation. For our purpose, the second point of view will be more useful. Having these properties in mind, one can show that

$$\widehat{\delta} \circ \mathfrak{F} = (\widehat{\delta} \circ \mathfrak{f} \wedge \mathfrak{F}) \circ \Delta$$

and

$$\widehat{[-,-]} \circ \mathfrak{F} = \left(\left(\widehat{[-,-]} \circ \mathfrak{f} + \frac{1}{2} \widehat{[-,-]} \circ (\mathfrak{f} \wedge \mathfrak{f}) \circ \Delta - ((\widehat{[-,-]} \circ \mathfrak{f}) \wedge \mathfrak{f}) \circ \Delta \right) \wedge \mathfrak{F} \right) \circ \Delta \, .$$

Besides the properties of $\widehat{\delta}$ and $\widehat{[-,-]}$, we also used cocommutativity and coassociativity of Δ. Thus we can equivalently define the QOCHA by

$$\begin{aligned}& \mathfrak{f} \circ \mathfrak{L}_c + \frac{\hbar}{2} \left(\mathfrak{f} \circ D(e_i) \wedge \mathfrak{f} \circ D(e^i) \right) \circ \Delta \\ & = \mathfrak{L}_o \circ \mathfrak{f} + \frac{1}{2} \widehat{[-,-]} \circ \left(\mathfrak{f} \wedge \mathfrak{f} \right) \circ \Delta - \left((\widehat{[-,-]} \circ \mathfrak{f}) \wedge \mathfrak{f} \right) \circ \Delta \, .\end{aligned} \tag{5.6}$$

The individual terms in Eq. (5.6) can be identified with the five distinct sewing operations of bordered Riemann surfaces with closed string insertions (punctures in the bulk) and open string insertions (punctures on the boundaries). The sewing joins either two open string insertions or two closed string insertions. In addition, the sewing may involve a single surface or two surfaces.

(1) Take an open string insertion of one surface and sew it with another open string insertion on a second surface. The genus of the resulting surface is the sum of the

genera of the individual surfaces, whereas the number of boundaries decreases by one. This operation is identified with

$$\frac{1}{2}\widehat{[-,-]} \circ (\mathfrak{f} \wedge \mathfrak{f}) \circ \Delta - \Big((\widehat{[-,-]} \circ \mathfrak{f}) \wedge \mathfrak{f} \Big) \circ \Delta \,.$$

(2) Sewing of two open string insertions living on the same boundary. This operation obviously increases the number of boundaries by one but leaves the genus unchanged. It is described by

$$\widehat{\delta} \circ \mathfrak{f} \,,$$

in the homotopy algebra.

(3) Consider a surface with more than one boundary. Take an open string insertion of one boundary and sew it with another open string insertion on a second boundary. This operation increases the genus by one and decreases the number of boundaries by one. It is identified with

$$\widehat{[-,-]} \circ \mathfrak{f} \,.$$

(4) Sewing of two closed string insertion, both lying on the same surface attaches a handle to the surface and hence increases the genus by one, whereas the number of boundaries does not change. We identify it with

$$\mathfrak{f} \circ D(\omega_c^{-1}) \,.$$

(5) Take a closed string insertion of one surface and sew it with another closed string insertion on a second surface. The genus and the number of boundaries of the resulting surface is the sum of the genera and the sum of the number of boundaries, respectively, of the input surfaces. The sewing in the case where both surfaces have open and closed insertions is identified with

$$\big(\mathfrak{f} \circ D(e_i) \wedge \mathfrak{f} \circ D(e^i) \big) \circ \Delta \,,$$

whereas the sewing involving a surface with closed string insertions only and another surface with open and closed string insertions is identified with

$$\mathfrak{f} \circ L_q \,.$$

The above analysis provides the geometric interpretation of all individual terms in (5.6).

Let us now focus on the vertices with open string insertions only. These vertices are also comprised in the IBL_∞-morphism $\mathfrak{F}$ and defined by setting the closed string inputs to zero. More precisely, let $\mathfrak{m} = \mathfrak{f}|$ be the restriction of $\mathfrak{f}$ onto the subspace without closed strings. The weighted sum of open string vertices is then given by

$$\mathfrak{m} = \sum_{b=1}^{\infty}\sum_{g=0}^{\infty} \hbar^{g+b-1} m^{b,g} , \qquad m^{b,g} \in \mathscr{A}_o^{\wedge b} ,$$

where $m^{b,g} = f^{b,g}|$. The complement of $\mathfrak{m}$—the vertices with at least one closed string input—is denoted by $\mathfrak{g}$, so that

$$\mathfrak{f} = \mathfrak{m} + \mathfrak{g} . \tag{5.7}$$

Remark 5.1 In the classical limit $\hbar \to 0$, we expect to recover the OCHA defined by Kajiura and Stasheff. Indeed, the IBL_∞-morphism $\mathfrak{F}$ reduces to an L_∞-morphism, the loop algebra $\mathfrak{L}_c$ of closed strings reduces to the L_∞-algebra $L_{cl} := L^0$ and the *IBL*-algebra on the space of cyclic coderivations becomes an ordinary Lie algebra. The defining Eq. (5.6) of the QOCHA simplifies to

$$f_{cl} \circ L_{cl} = \frac{1}{2}[f_{cl}, f_{cl}] \circ \Delta , \tag{5.8}$$

where $f_{cl} := f^{1,0}$ is the component of $\mathfrak{f}$ with one boundary and genus zero and the corresponding L_∞-morphism is given by $\sum_n \frac{1}{n!} f_{cl}{}^{\wedge n} \circ \Delta_n$ (see Sect. A.2). Separating the purely open string vertices m_{cl} from f_{cl}, we see that those have to satisfy the axioms of an A_∞-algebra (since $L_{cl}| = 0$), i.e., they define a classical open string field theory. Thus, the space $\mathscr{A}_o$ turns into a dgla with differential $d_H = [m_{cl}, -]$ and Eq. (5.8) finally reads

$$n_{cl} \circ L_{cl} = d_H \circ n_{cl} + \frac{1}{2}[n_{cl}, n_{cl}] \circ \Delta , \tag{5.9}$$

where $n_{cl} = f_{cl} - m_{cl} : TV_c \to \mathscr{A}_o$ denotes the vertices with at least one closed string input and one non-empty boundary.

Similarly, we define $\mathfrak{n} = \mathfrak{f} - \mathfrak{m}$ and the QOCHA in terms of $\mathfrak{n}$ reads

$$\mathfrak{N} \circ \mathfrak{L}_c = \mathfrak{L}_o' \circ \mathfrak{N} , \tag{5.10}$$

where $\mathfrak{N} = \sum_{n=0}^{\infty} \frac{1}{n!} \mathfrak{n}^{\wedge n} \circ \Delta_n$ and $\mathfrak{L}_o' = \widehat{d_H} + \mathfrak{L}_o$.

Equation (5.9) is precisely the OCHA. The physical interpretation of n_{cl} is that it describes the deformation of open string field theory by turning on a closed string background. The vanishing of the right-hand side is the condition for a consistent classical field theory of open strings, while the left-hand side vanishes if the closed string background solves the classical closed string field theory equations of motion. Equation (5.9) then implies that the open-closed vertices define a consistent classical open string field theory if the closed string background satisfies the classical closed string equations of motion. The inverse assertion does not follow from (5.9). However, it holds true for infinitesimal closed string deformations. More precisely, upon linearizing equation (5.9) in $c \in V_c$ we get

$$n_{cl}(L_{cl}(c)) = d_H(n_{cl}(c)) \,, \tag{5.11}$$

where $L_{cl} \in \mathrm{Coder}^{cycl}(TV_c)$ is determined by $l_{cl} = \pi_1 \circ L_{cl} \in \mathrm{Hom}^{cycl}(TV_c)$, the closed string vertices of genus zero (see Sect. A.2). In string field theory, the vertex with just one input $(l_{cl})_1$ is the closed string BRST operator Q_c. Thus Eq. (5.11) is equivalent to

$$n_{cl}(Q_c(c)) = d_H(n_{cl}(c)) \,,$$

that is, $n_{cl} \circ i_1$ induces a chain map from the BRST complex of closed strings to the cyclic Hochschild complex of open strings. The cohomology of Q_c (BRST cohomology) defines the space of physical states, whereas the cohomology of d_H (cyclic cohomology) characterizes the infinitesimal deformations of the initial open string field theory m_{cl} as discussed in Sect. 4.2. There we have seen that the BRST cohomology of closed strings is indeed isomorphic to the cyclic Hochschild cohomology of open strings.

5.3 Summary, Comment, and Remarks Towards Part II

The detailed discussion of the open-closed SFT in Sect. 5.2 is taken from the point of view of IBL_∞-algebras and their morphisms. Nevertheless, the description of geometric vertices $\nu_{n,m}^{b,g}$ as well of the open-closed SFT action in (5.1) can directly be interpreted in the spirit of Fig. 1.1. As already noticed, this construction would be a rather straightforward combination of the open and closed theories.

The starting point would be the open-closed modular operad. Since we are not going to describe it in full detail, we at least indicate its nature here. It is a two-colored operad with colors corresponding to open and closed strings (punctures), respectively. In terms of corollas used in Part II, we would have to consider corollas with two kinds of legs. This can be pictured informally as follows. Think about bordered Riemann surfaces with punctures both on the boundaries (open ones) as well as in the bulk (closed ones). We allow also boundaries with no punctures. Now, forget about all the structure but genus, number of boundaries, number of closed punctures, and the distribution of punctures on the boundaries. What we have are corollas remembering the genus and having their legs colored as open or closed, the open ones grouped together accordingly to their respective boundaries. There is an obvious two-fold action of permutations, we can either permute closed punctures between themselves or open punctures between themselves. We allow for permuting of punctures between different boundaries, but we do not allow for permutations mixing up mutually closed and open punctures. Also operations sewing together open and closed punctures are forbidden. Sewing and self-sewing operations within the two separate sectors remain the same as described before.

The open-closed geometric vertices $v_{n,m}^{b,g}$ satisfying the (geometric) quantum BV master equation result from a decomposition of the moduli space of bordered Riemann surfaces with punctures both in the bulk as well as on the boundary components. Again, it can be understood as an odd modular operad morphism going from the (two-colored version of) Feynman transform of the above described open-closed modular operad to the moduli space operad, cf. discussion in Sects. 3.8 and 4.3. The latter one is the odd modular operad on singular chain complex with operations induced from properly defined sewing/self-sewing of the Riemann surfaces. As briefly mentioned in Sect. 5.1, the decomposition of the moduli spaces comes as a solution to the corresponding minimal area problem. This is a rather informal description of the horizontal morphism of Fig. 1.1 in the present situation.

The vertical arrow in Fig. 1.1 is the morphism from the odd moduli space operad to the open-closed endomorphism operad provided by the open-closed CFT. Recall that the CFT state space $V = V_o \oplus V_c$ is equipped with the odd symplectic form $\omega_o + \omega_c$. The morphism is described similarly as in the closed case, cf. (3.22). The resulting action (5.1) is the generating function of the quantum open-closed operations $f_{n,m}^{b,g}$, these are graded symmetric functions of closed states. The graded symmetry with respect to the open states holds for cyclic permutations within a boundary component and for permutations of the whole boundary components.

Recall, cf. Sect. 3.8, that the main difference of the IBL_∞ interpretation of the closed SFT algebra, as opposed to the loop homotopy algebra, was that we considered, together with the n-ary brackets l_n^g, also the BV operator Δ as a special cobracket with zero inputs and two outputs. The resulting structure is naturally described in the dual description by the nilpotent full BV operator $\hbar\Delta + \{S, -\}$. Obviously, starting from a quantum A_∞-algebra we could do the same. However, this is not what is used for the interpretation/description of the quantum-open homotopy algebra in Sect. 5.2. There, the starting point for the description of the open sector is the cyclic Hochschild complex equipped with an *IBL*-algebra structure. Roughly speaking, now we think of the inputs/outputs of interactions being disc with open string insertions, not the open strings themselves. So "sewing" two discs with open string insertions using the odd symplectic form gives the multiplication, whereas the self-sewing gives the comultiplication. The algebraic vertices are formally split into two groups, the ones corresponding to closed strings exclusively and the remaining ones. The latter ones correspond to an IBL_∞-morphism form the IBL_∞-algebra corresponding to closed strings to the *IBL*-algebra on the cyclic Hochschild complex of the open sector. The corresponding mathematical structures are elucidated in the following appendix to Part I and in Part II, Sect. 8.3.

Further Reading

For a detailed derivation of the open-closed BV bracket and the corresponding BV action see

- B. Zwiebach, "Oriented open-closed string theory revisited", Annals Phys. 267, (1998) 193.

The classical open-closed homotopy algebra was derived in

- H. Kajiura, J. Stasheff, "Open-closed homotopy algebra in mathematical physics", J. Math. Phys. 47, 023506 (2006).

Our description of the quantum open-closed homotopy algebra in this chapter is based on

- K. Münster and I. Sachs, "Quantum Open-Closed Homotopy Algebra and String Field Theory", Commun. Math. Phys. 321 (2013) 769.

For a detailed description of the IBL_∞ algebra underlying quantum closed string theory see

- K. Cieliebak, K. Fukaya and J. Latschev, "Homological algebra related to surfaces with boundary", arXiv:1508.02741.

A detailed description of the open-closed modular operad can be found in

- M. Doubek and M. Markl, "Open-closed modular operads, the Cardy condition and string field theory", J. Noncommut. Geom. 12(4)(2018), 1359–1424.

The quantum open-closed homotopy algebra and its operadic origin are discussed in

- M. Doubek, B. Jurčo and K. Münster, "Modular operads and the quantum open-closed homotopy algebra", JHEP 2015(12), 1–55.

A_∞- and L_∞-Algebras

A

We review definitions of A_∞- and L_∞-algebras. In the following $A = \bigoplus_{n\in\mathbb{Z}} A_n$ will denote a graded vector space over some field $\Bbbk$ of characteristic 0 (more generally we could consider a module A over some commutative ring R with unit containing rational numbers). We will use the Koszul sign convention, that is, we generate a sign $(-1)^{xy}$ whenever we permute two objects x and y with their respective degrees denoted by the same symbols. If we permute several object, we abbreviate the Koszul sign by $(-1)^{\epsilon}$. To simplify the exposition, we will assume in this appendix all homogeneous pieces of the underlying graded vector spaces to be finite-dimensional.

A.1 A_∞-Algebras

Let us consider the tensor algebra of A

$$TA = \bigoplus_{n=0}^{\infty} A^{\otimes n} ,$$

and the comultiplication $\Delta : TA \to TA \otimes TA$ defined by

$$\Delta(a_1 \otimes \cdots \otimes a_n) = \sum_{i=0}^{n} (a_1 \otimes \cdots \otimes a_i) \otimes (a_{i+1} \otimes \cdots \otimes a_n) .$$

The comultiplication Δ makes TA a coassociative coalgebra, i.e.,

$$(\Delta \otimes \mathbb{1}) \circ \Delta = (\mathbb{1} \otimes \Delta) \circ \Delta .$$

© Springer Nature Switzerland AG 2020

M. Doubek et al., *Algebraic Structure of String Field Theory*, Lecture Notes in Physics 973, https://doi.org/10.1007/978-3-030-53056-3

In addition we have the obvious canonical projection maps $\pi_n : TA \to A^{\otimes n}$ and the inclusion maps $i_n : A^{\otimes n} \to TA$. A coderivation $D \in \mathrm{Coder}(TA)$ is a linear map having the property

$$(D \otimes \mathbb{1} + \mathbb{1} \otimes D) \circ \Delta = \Delta \circ D \,. \tag{A.1}$$

The defining property (A.1) implies that a coderivation $D \in \mathrm{Coder}(TA)$ is uniquely determined by a map $d \in \mathrm{Hom}(TA, A)$, i.e., $\mathrm{Coder}(TA) \cong \mathrm{Hom}(TA, A)$. Explicitly the correspondence reads

$$D \circ i_n = \sum_{i+j+k=n} \mathbb{1}^{\otimes i} \otimes d_j \otimes \mathbb{1}^{\otimes k} \,,$$

where $d_n := d \circ i_n$, $\mathbb{1}$ denotes the identity map on A and $d = \pi_1 \circ D$. The space of coderivations $\mathrm{Coder}(TA)$ turns out to be a Lie algebra with the Lie bracket defined by

$$[D_1, D_2] := D_1 \circ D_2 - (-1)^{D_1 D_2} D_2 \circ D_1 \,.$$

An A_∞-algebra is determined by a coderivation $M \in \mathrm{Coder}(TA)$ of degree 1 (degree -1 is considered if m_1 is supposed to be a boundary operator rather than a coboundary operator) that squares to zero,

$$M^2 = \frac{1}{2}[M, M] = 0 \qquad \text{and} \qquad |M| = 1 \,.$$

The corresponding homomorphism is defined by $m := \pi_1 \circ M$.

In the case where only m_1 and m_2 are non-vanishing, we recover the definition of a differential graded associative algebra up to a shift: Take $A[1] = \uparrow A$ to be the graded vector space defined by $(\uparrow A)_i = A_{i-1}$. One has the map $\uparrow : A \to \uparrow A$ whose only effect is increasing the degree by 1. Likewise, its inverse map $\downarrow : \uparrow A \to A$ decreases the degree by one. The maps corresponding to the shifted space $\uparrow A$ are defined by

$$\tilde{m}_n := \uparrow \circ m_n \circ (\downarrow)^{\otimes n} : (\uparrow A)^{\otimes n} \to \uparrow A \,.$$

The operations $\tilde{m}_1$ and $\tilde{m}_2$ then determine a differential graded associative algebra, if $m_n = 0$ for $n \geq 3$.

Consider now two A_∞-algebras (A', M') and (A'', M''). An A_∞-morphism $F \in \mathrm{Morph}(A', A'')$ from (A', M') to (A'', M'') is a degree 0 linear map $F : TA' \to TA''$ satisfying

$$\Delta \circ F = (F \otimes F) \circ \Delta \,, \qquad F \circ M' = M'' \circ F. \tag{A.2}$$

The first equation in (A.2) implies that a morphism $F \in \text{Morph}(A', A'')$ is determined by a map $f \in \text{Hom}(TA', A'')$. The explicit relation reads

$$F = \sum_{n=0}^{\infty} f^{\otimes n} \circ \Delta_n \,,$$

where $\Delta_n : TA' \to TA'^{\otimes n}$ denotes the $(n-1)$-fold comultiplication and $f = \pi_1 \circ F$. We use the convention that $\Delta_1 := \mathbb{1}$ and that Δ_0 equals the unit in the field $\Bbbk$. An important property is that the composition of two A_∞-morphisms is again an A_∞-morphism, i.e., for $F \in \text{Morph}(A', A'')$ and $G \in \text{Morph}(A'', A''')$, $G \circ F \in \text{Morph}(A', A''')$. This is a direct consequence of Eq. (A.2).

The concept of Maurer–Cartan elements of A_∞-algebras is closely related to that of A_∞-morphisms. We define the exponential in the completion $\hat{T}A$ of TA as

$$e^a := \sum_{n=0}^{\infty} a^{\otimes n} \,.$$

A Maurer–Cartan element $a \in A$ of an A_∞-algebra (A, M) is a degree zero element that satisfies

$$M(e^a) = 0 \qquad \Leftrightarrow \qquad \sum_{n=0}^{\infty} m_n(a^{\otimes n}) = 0 \,.$$

Note that $\Delta(e^a) = e^a \otimes e^a$. Thus we can interpret the exponential e^a of a Maurer–Cartan element $a \in A$ as a constant morphism from the trivial A_∞ algebra to (A, M), that is, $f_0 = a$ and $f_n = 0$ for all $n \geq 1$. Since we know that the composition of two A_∞-morphisms is again an A_∞-morphism and that a Maurer–Cartan element can be interpreted as a constant A_∞-morphism, it follows that an A_∞-morphism sends Maurer–Cartan elements into Maurer–Cartan elements. The same statement is true for L_∞-algebras (see Sect. A.2).

The language of coderivations is also useful for describing deformations of A_∞-algebras. Deformations of an A_∞-algebra (A, M) are controlled by the differential graded Lie algebra $\text{Coder}(TA)$ with differential $d_h := [M, -]$ and bracket $[-, -]$. Since $\text{Coder}(TA) \cong \text{Hom}(TA, A)$, d_h and $[-, -]$ have their counterparts defined on $\text{Hom}(TA, A)$, namely the Hochschild differential and the Gerstenhaber bracket. An infinitesimal deformation of an A_∞-algebra is characterized by the Hochschild cohomology $H^1(d_h, \text{Coder}(TA))$, i.e., the cohomology of d_h at degree 1. A deformation of an A_∞-algebra is an element $D \in \text{Coder}(TA)$ of degree 1 that satisfies the Maurer–Cartan equation

$$d_h(D) + \frac{1}{2}[D, D] = 0 \qquad \Leftrightarrow \qquad (M + D)^2 = 0 \,.$$

We will need one more concept in the context of A_∞-algebras which is called the cyclicity. Assume that A is an A_∞-algebra whose underlying graded vector space is additionally endowed with an odd symplectic structure $\omega : A \otimes A \to \Bbbk$ of degree -1. We call $d \in \text{Hom}(TA, A)$ cyclic if the multilinear map

$$\omega(d\,, -) : TA \to \Bbbk$$

is cyclically symmetric, i.e.,

$$\omega(d_n(a_1, \dots, a_n), a_{n+1}) = (-1)^\epsilon \omega(d_n(a_2, \dots, a_{n+1}), a_1)\ .$$

Since we have the notion of cyclicity for $\text{Hom}(TA, A)$, we also have the notion of cyclicity for $\text{Coder}(TA)$ due to the isomorphism $\text{Coder}(TA) \cong \text{Hom}(TA, A)$. We denote the space of cyclic coderivations by $\text{Coder}^{cycl}(TA)$. An A_∞-algebra (A, M, ω) is called a cyclic A_∞-algebra if $M \in \text{Coder}^{cycl}(TA)$. It is straightforward to prove that $\text{Coder}^{cycl}(TA)$ is closed with respect to the Lie bracket $[-, -]$, and thus we can consider deformations of cyclic A_∞-algebras which are controlled by the differential graded Lie algebra $\text{Coder}^{cycl}(TA)$. The cohomology $H(d_h, \text{Coder}^{cycl}(TA))$ is called the cyclic cohomology.

A.2 L_∞-Algebras

Many of the constructions in the context of L_∞-algebras are analogous to those of A_∞-algebras. The main difference is that the definition of an L_∞-algebra is based on the graded symmetric algebra SA instead of the tensor algebra TA. The graded symmetric algebra SA is defined as the quotient TA/I, where I denotes the two-sided ideal generated by the elements

$$c_1 \otimes c_2 - (-1)^{c_1 c_2} c_2 \otimes c_1, \quad c_1, c_2 \in A.$$

The product $\otimes$ defined in TA induces the graded symmetric product $\wedge$ in SA. The symmetric algebra is the direct sum of the symmetric powers of A,

$$SA = \bigoplus_{n=0}^{\infty} A^{\wedge n}\ .$$

All that is simply saying that an element $c_1 \wedge \cdots \wedge c_n \in A^{\wedge n}$ is graded symmetric, that is

$$c_{\sigma_1} \wedge \cdots \wedge c_{\sigma_n} = (-1)^\epsilon c_1 \wedge \cdots \wedge c_n$$

for any permutation $\sigma \in \Sigma_n$, where Σ_n denotes the permutation group of n elements. Here, the elements c_i are assumed to be of a definite degree and ϵ is the Koszul sign. The comultiplication $\Delta : SA \to SA \otimes SA$ is defined by

$$\Delta(c_1, \cdots, c_n) = \sum_{i=0}^{n} \sum_{\sigma}{}' (c_{\sigma_1} \wedge \cdots \wedge c_{\sigma_i}) \otimes (c_{\sigma_{i+1}} \wedge \cdots \wedge c_{\sigma_n}) ,$$

where $\sum'_\sigma$ indicates the sum over all permutations $\sigma \in \Sigma_n$ constrained to

$$\sigma_1 < \cdots < \sigma_i \quad \text{and} \quad \sigma_{i+1} < \cdots < \sigma_n .$$

A coderivation $D \in \mathrm{Coder}(SA)$ is a linear map satisfying

$$(D \otimes \mathbb{1} + \mathbb{1} \otimes D) \circ \Delta = \Delta \circ D . \tag{A.3}$$

Again, the isomorphism $\mathrm{Coder}(SA) \cong \mathrm{Hom}(SA, A)$ holds. The correspondence between a coderivation $D \in \mathrm{Coder}(SA)$ and its associated map $d = \pi_1 \circ D \in \mathrm{Hom}(SA, A)$ is given by

$$D \circ i_n = \sum_{i+j=n} \sum_{\sigma}{}' (d_i \wedge \mathbb{1}^{\wedge j}) \circ \sigma ,$$

where σ in the right-hand side denotes the map that sends $c_1 \wedge \cdots \wedge c_n$ into $(-1)^\epsilon c_{\sigma_1} \wedge \cdots \wedge c_{\sigma_n}$ (again $d_n = d \circ i_n$ and $\mathbb{1}$ is the identity map on A).

An L_∞-algebra is determined by a coderivation $L \in \mathrm{Coder}(SA)$ of degree 1 that squares to zero,

$$L^2 = 0 \qquad \text{and} \qquad |L| = 1 .$$

An L_∞-morphism $F \in \mathrm{Morph}(A', A'')$ from an L_∞-algebra (A', L') to another L_∞-algebra (A'', L'') is a degree 0 linear map $F : SA \to SA$ such that

$$\Delta \circ F = (F \otimes F) \circ \Delta , \qquad F \circ L' = L'' \circ F. \tag{A.4}$$

Such an F is determined by a map $f = \pi_1 \circ F \in \mathrm{Hom}(SA, A')$ through

$$F = \sum_{n=0}^{\infty} \frac{1}{n!} f^{\wedge n} \circ \Delta_n , \tag{A.5}$$

where $\Delta_n : SA \to (SA)^{\otimes n}$ denotes the $(n-1)$-fold comultiplication.

Analogously to A_∞-algebras, a Maurer–Cartan element $c \in A$ of an L_∞-algebra (A, L) is essentially a constant morphism from the trivial L_∞-algebra $\Bbbk$ sending $1 \in \Bbbk$ to c, that is,

$$L(e^c) = 0 \qquad \text{and} \qquad |c| = 0\,,$$

where the exponential, in the completion of the symmetric algebra, is given by

$$e^c = \sum_{n=0}^{\infty} \frac{1}{n!} c^{\wedge n}$$

and satisfies $\Delta(e^c) = e^c \otimes e^c$.

Finally, there is also the notion of cyclicity in the context of L_∞-algebras. Let (A, L) be a L_∞-algebra whose underlying vector space is equipped with an odd symplectic structure ω of degree -1. We call a coderivation $D \in \mathrm{Coder}(SA)$ cyclic if the corresponding multilinear map $\omega(d, -)$ is graded symmetric, i.e.,

$$\omega(d_n(c_{\sigma_1}, \ldots, c_{\sigma_n}), c_{\sigma_{n+1}}) = (-1)^{\epsilon} \omega(d_n(c_1, \ldots, c_n), c_{n+1})\,.$$

We denote the space of cyclic coderivations by $\mathrm{Coder}^{cycl}(SA)$.

As a simple illustration of L_∞-morphisms we give a background shift in closed string field theory. Consider the classical action of closed string field theory, the theory with genus zero vertices l_{cl} only. The corresponding coderivation L_{cl} defines an L_∞-algebra and the action reads

$$S_{c,cl} = \omega_c(l_{cl}, -)(e^c)\,.$$

Shifting the background simply means that we expand the string field c around c' rather than around zero. The action in the new background is $\omega_c(l_{cl}, -)(e^{c'+c})$. Hence, the vertices $l_{cl}[c']$ in the shifted background read

$$l_{cl}[c'] = l_{cl} \circ E(c')\,,$$

where $E(c')$ is the map defined by

$$E(c')(c_1 \wedge \cdots \wedge c_n) = e^{c'} \wedge c_1 \wedge \cdots \wedge c_n\,.$$

In the language of homotopy algebras, this shift is implemented by

$$L_{cl}[c'] = E(-c') \circ L_{cl} \circ E(c')\,.$$

Obviously, $E(-c')$ is the inverse map of $E(c')$. Furthermore,

$$\Delta \circ E(c') = E(c') \otimes E(c')$$

and therefore $L_{cl}[c']$ defines also an L_∞-algebra. Thus, $E(c')$ is an L_∞-morphism. There is a subtlety if the new background does not satisfy the field equations. The initial L_∞-algebra is determined by the vertices $(l_{cl})_n$ where there is no vertex for $n = 0$, i.e., $(l_{cl})_0 = 0$. A non-vanishing $(l_{cl})_0$ would correspond to a term in the action that depends linearly on the field. In the new background we get

$$(l_{cl}[c'])_0 = \sum_{n=0}^{\infty} \frac{1}{n!}(l_{cl})_n(c'^{\wedge n}) ,$$

and thus the L_∞-algebra $L_{cl}[c']$ has no $n = 0$ vertex only if c' satisfies the field equations. In general, the new l_1 operation is no longer a differential of A, instead we have $l_1 \circ l_1 + l_2 \circ (l_0 \wedge \mathbb{1}) = 0$, i.e., we have a curved L_∞-algebra.

Homotopy Involutive Lie Bialgebras

B

Homotopy algebras as reviewed in Appendix A are suitable for describing the algebraic structures of classical open-closed string field theory. If one tries to describe quantum open-closed string field theory—with the set of vertices satisfying the full quantum BV master equation—in the framework of homotopy algebras, the appropriate language is that of homotopy involutive Lie bialgebras, or IBL_∞-algebras.[1] An IBL_∞-algebra is a generalization of an L_∞-algebra. Its axioms are formulated in terms of higher order coderivations—a concept that will be introduced in the next section—and requires an auxiliary parameter $x \in \Bbbk$ (later on we will identify that parameter with $\hbar$). We will also recall the notion of morphisms and Maurer–Cartan elements in the context of IBL_∞-algebras. Our exposition is based on work of Cieliebak, Fukaya, and Latschev cited at the end of Chap. 5. In the following, we collect their results (in a slightly different notation) to make our exposition self-contained. An alternative description is provided by Sect. 8.3.

B.1 Higher Order Coderivations

We already know what a coderivation (of order one) on SA is, cf. Eq. (A.3). We defined it by an algebraic equation involving the comultiplication Δ. The essence of that equation was that a coderivation $D \in \mathrm{Coder}(SA)$ was uniquely determined by a homomorphism $d \in \mathrm{Hom}(SA, A)$. Explicitly we had

$$D \circ i_n = \sum_{i+j=n} \sum_{\sigma}{}' (d_i \wedge \mathbb{1}^{\wedge j}) \circ \sigma\ , \tag{B.1}$$

where $\pi_1 \circ D = d$.

[1] As we already mentioned, an alternative description using the language of quantum open-closed homotopy algebra is also possible.

© Springer Nature Switzerland AG 2020

M. Doubek et al., *Algebraic Structure of String Field Theory*, Lecture Notes in Physics 973, https://doi.org/10.1007/978-3-030-53056-3

There are two ways to define higher order coderivations. One is based on algebraic relations like that in Eq. (A.3). A coderivation of order two is, for example, characterized by

$$\Delta_3 \circ D - \sum_{\sigma}{}' \sigma \circ (\Delta \circ D \otimes \mathbb{1}) \circ \Delta + \sum_{\sigma}{}' \sigma \circ (D \otimes \mathbb{1}^{\otimes 2}) \circ \Delta_3 = 0\,,$$

where $\sum_{\sigma}'$ denotes the sum over inequivalently acting permutations in Σ_3, the permutation group of three elements, and $\sigma : SA^{\otimes 3} \to SA^{\otimes 3}$ is the map that permutes the three factors. For completeness we state an algebraic definition of a coderivation $D \in \mathrm{Coder}^n(SA)$ of order n,

$$\sum_{i=0}^{n} \sum_{\sigma}{}' (-1)^i \sigma \circ (\Delta_{n+1-i} \circ D \otimes \mathbb{1}^{\otimes i}) \circ \Delta_{i+1} = 0\,. \tag{B.2}$$

As in the case of a coderivation of order one, this relation is saying—and this is an alternative definition of higher order coderivations—that a coderivation $D \in \mathrm{Coder}^n(SA)$ of order n is uniquely determined by a map $d \in \mathrm{Hom}(SA, \Sigma^n A)$, where $\Sigma^n A = \oplus_{i=0}^{n} A^{\wedge i}$. Thus in contrast to a coderivation of order one, a coderivation of order n is determined by a linear map on SA with n and less outputs rather than just one output. The explicit relation between $D \in \mathrm{Coder}^n(SA)$ and $d \in \mathrm{Hom}(SA, \Sigma^n A)$ is

$$D \circ i_n = \sum_{i+j=n} \sum_{\sigma}{}' (d_i \wedge \mathbb{1}^{\wedge j}) \circ \sigma\,,$$

which is a naive generalization of Eq. (B.1).

A trivial observation is that a coderivation of order $n-1$ is also a coderivation of order n, by simply defining the component with n outputs to be zero, that is,

$$\mathrm{Coder}^{n-1}(SA) \subset \mathrm{Coder}^n(SA)\,.$$

We call a coderivation $D \in \mathrm{Coder}^n(SA)$ of order n a strict coderivation of order n if the corresponding map d is in $\mathrm{Hom}(SA, A^{\wedge n})$, that is, if the map d has exactly n outputs. In that case we can identify $d = \pi_n \circ D$.

Next recall the graded commutator

$$[D_1, D_2] = D_1 \circ D_2 - (-1)^{D_1 D_2} D_2 \circ D_1\,,$$

where D_1, D_2 are arbitrary higher order coderivations. Using the defining equations (B.2), it can be shown that

$$[\mathrm{Coder}^i(SA), \mathrm{Coder}^j(SA)] \subset \mathrm{Coder}^{i+j-1}(SA)\,. \tag{B.3}$$

In the case $i = j = 1$ we recover that $[-,-]$ defines a Lie algebra on $\mathrm{Coder}^1(SA)$, but we see that $[-,-]$ does not define a Lie algebra at higher orders $n > 1$. Of course, we can make the collection of all higher order coderivations a Lie algebra, but in the next section we will see that there is still a finer structure.

B.2 *IBL_∞*-Algebras

Now we have all tools to define IBL_∞-algebras. We will furthermore see that one recovers an involutive Lie bialgebra (IBL-algebra) as a special case of an IBL_∞-algebra. Consider the space

$$\mathrm{Coder}(SA, x) := \bigoplus_{n=1}^{\infty} x^{n-1}\mathrm{Coder}^n(SA) ,$$

where $x \in \Bbbk$ is some auxiliary parameter. An element $\mathfrak{D} \in \mathrm{Coder}(SA, x)$ can be expanded as

$$\mathfrak{D} = \sum_{n=1}^{\infty} x^{n-1} D^{(n)} ,$$

where $D^{(n)} \in \mathrm{Coder}^n(SA)$. In the following, we will indicate coderivations of order n by the superscript (n) and strict coderivations of order n by the superscript n. We can decompose every coderivation of order n into strict coderivations of order smaller than or equal to n. Accordingly, we denote the strict coderivation of order $n-g$ corresponding to a coderivation $D^{(n)}$ of order n by $D^{n-g,g}$, $g \in \{0, \ldots, n-1\}$ (in the main text g was identified as the genus). Thus, we have

$$D^{(n)} = \sum_{g=0}^{n-1} D^{n-g,g} ,$$

and $\mathfrak{D}$ expressed in terms of strict coderivations reads

$$\mathfrak{D} = \sum_{n=1}^{\infty}\sum_{g=0}^{\infty} x^{n+g-1} D^{n,g} .$$

Due to Eq. (B.3), we have

$$[\mathfrak{D}_1, \mathfrak{D}_2] \in \mathrm{Coder}(SA, x) ,$$

that is, the commutator $[-,-]$ turns $\mathrm{Coder}(SA, x)$ into a graded Lie algebra. The space $\mathrm{Coder}(SA, x)$ is the Lie algebra on which our definition of IBL_∞-algebras is based. From a conceptual point of view, nothing new happens in the definition of IBL_∞-algebras when compared to the one of L_∞- and A_∞-algebras. The difference is essentially that the underlying objects are more complicated. An IBL_∞-algebra is defined by an element $\mathfrak{L} \in \mathrm{Coder}(SA, x)$ of degree 1 that squares to zero:

$$\mathfrak{L}^2 = 0 \qquad \text{and} \qquad |\mathfrak{L}| = 1\,.$$

For completeness, we will now describe IBL-algebras as a special case of IBL_∞-algebras. Consider an element $\mathfrak{L} \in \mathrm{Coder}(SA, x)$ that consists of a strict coderivation of order one and a strict coderivation of order two only:

$$\mathfrak{L} = L^{1,0} + xL^{2,0}\,.$$

Furthermore, we restrict to the case where the only non-vanishing components of $l^{1,0} := \pi_1 \circ L^{1,0} : SA \to A$ and $l^{2,0} := \pi_2 \circ L^{2,0} : SA \to A^{\wedge 2}$ are

$$d := l^{1,0} \circ i_1 : A \to A\,, \quad [-,-] := l^{1,0} \circ i_2 : A^{\wedge 2} \to A\,, \quad \delta := l^{2,0} \circ i_1 : A \to A^{\wedge 2}\,.$$

To recover the definition of an involutive Lie bialgebra, we have to shift the degree by one (see Appendix A), i.e., we define the operations on the shifted space $\uparrow A$ by

$$\widetilde{d} := \uparrow \circ d \circ \downarrow\,, \qquad \widetilde{[-,-]} := \uparrow \circ [-,-] \circ (\downarrow)^{\wedge 2}\,, \quad \widetilde{\delta} := \uparrow^{\wedge 2} \circ\, \delta \circ \downarrow\,.$$

The requirement $\mathfrak{L}^2 = 0$ is then equivalent to the following seven conditions

$$\widetilde{d}^2 = 0, \tag{B.4}$$

$$\widetilde{d}\,\widetilde{[-,-]} + \widetilde{[-,-]}\,(\widetilde{d} \wedge \mathbb{1} + \mathbb{1} \wedge \widetilde{d}) = 0, \tag{B.5}$$

$$(\widetilde{d} \wedge \mathbb{1} + \mathbb{1} \wedge \widetilde{d})\,\widetilde{\delta} + \widetilde{\delta}\,\widetilde{d} = 0, \tag{B.6}$$

$$\textstyle\sum'_\sigma \widetilde{[-,-]}\,(\widetilde{[-,-]} \wedge \mathbb{1})\,\sigma = 0, \tag{B.7}$$

$$\textstyle\sum'_\sigma \sigma\,(\widetilde{\delta} \wedge \mathbb{1} + \mathbb{1} \wedge \widetilde{\delta})\,\widetilde{\delta} = 0, \tag{B.8}$$

$$\textstyle\sum'_\sigma (\widetilde{[-,-]} \wedge \mathbb{1})\,\sigma\,(\widetilde{\delta} \wedge \mathbb{1} + \mathbb{1} \wedge \widetilde{\delta}) + \widetilde{\delta}\,\widetilde{[-,-]} = 0, \text{ and} \tag{B.9}$$

$$\widetilde{[-,-]}\,\widetilde{\delta} = 0. \tag{B.10}$$

In the above display, (B.4) means that $\widetilde{d}$ is a differential, (B.5) that $\widetilde{d}$ is a derivation for $\widetilde{[-,-]}$, (B.6) that $\widetilde{d}$ is a derivation for $\widetilde{\delta}$, (B.7) is the Jacobi identity for $\widetilde{[-,-]}$, (B.8) is the co-Jacobi identity for $\widetilde{\delta}$, (B.9) is the compatibility between $\widetilde{\delta}$ and $\widetilde{[-,-]}$, and (B.10) is the involutivity. We recognize the axioms defining a differential involutive Lie bialgebra.

B.3 IBL_∞-Morphisms and Maurer–Cartan Elements

An L_∞-morphism was defined by two equations (A.4). The first one involves the comultiplication and implies that an L_∞-morphism can be expressed by a linear map from SA to A, cf. (A.5). We do not know of a suitable generalization of the first equation in (A.4) to the case of IBL_∞-algebras, but instead one can easily generalize Eq. (A.5). The second equation of (A.4) is just saying that the morphism commutes with the differentials and looks identically in the case of IBL_∞-algebras.

Let $(A', \mathfrak{L}')$, $(A'', \mathfrak{L}'')$ be IBL_∞-algebras. An IBL_∞-morphism $\mathfrak{F} \in \mathrm{Morph}(A', A'', x)$ is defined by

$$\mathfrak{F} = \sum_{n=0}^{\infty} \frac{1}{n!} \mathfrak{f}^{\wedge n} \circ \Delta_n \,, \qquad \mathfrak{F} \circ \mathfrak{L}' = \mathfrak{L}'' \circ \mathfrak{F} \qquad \text{and} \qquad |\mathfrak{F}| = 0 \,, \tag{B.11}$$

where

$$\mathfrak{f} = \sum_{n=0}^{\infty} x^{n-1} f^{(n)} \qquad \text{and} \qquad f^{(n)} : SA' \to \Sigma^n A'' \,.$$

The precise meaning of the morphism $\mathfrak{F}$ can be found in Part II, Sect. 8.3, especially Lemma 8.2. Recall that $\Sigma^n A'' = \oplus_{i=1}^{n} A''^{\wedge i}$. We can therefore decompose $f^{(n)}$ into a set of maps $f^{n-g,g} : SA' \to A''^{\wedge n-g}$, $g \in \{0, \ldots, n-1\}$ in the same way as we decomposed higher order coderivations. Expressed in terms of maps $f^{n,g}$ we have

$$\mathfrak{f} = \sum_{n=1}^{\infty} \sum_{g=0}^{\infty} x^{n+g-1} f^{n,g} \,. \tag{B.12}$$

Due to the lack of an algebraic relation governing the structure of an IBL_∞-morphism—an equation generalizing the first equation in (A.4)—it is not obvious that the composition of two morphisms yields again a morphism. Nevertheless, this can be shown to be true.

To complete this section, we finally state what a Maurer–Cartan element of an IBL_∞-algebra $(A, \mathfrak{L})$ is. Let $c^{n,g} \in A^{\wedge n}$ be of degree zero. The expression $\mathbf{c} = \sum_{n=1}^{\infty} \sum_{g=0}^{\infty} x^{n+g-1} c^{n,g}$ is called a Maurer–Cartan element of $(A, \mathfrak{L})$ if

$$\mathfrak{L}(e^{\mathbf{c}}) = 0 \,.$$

Again we can interpret a Maurer–Cartan element as a constant morphism from the trivial IBL_∞-algebra to $(A, \mathfrak{L})$. Here, the exponential is defined in the same way as in the case of L_∞-algebras, i.e., $e^{\mathbf{c}} = \sum_{n=0}^{\infty} \frac{1}{n!} \mathbf{c}^{\wedge n}$, now being a formal power series with values in the completion of the symmetric algebra over A.

Part II

Mathematical Interpretation

Conventions Used in This Part

If not stated otherwise, all algebraic objects will be considered over a fixed field $\mathbb{k}$ of characteristic zero. The symbol $\otimes$ will be reserved for the tensor product over $\mathbb{k}$. Given a set S, Span(S) will denote the $\mathbb{k}$-vector space generated by S. We will denote by $\mathbb{1}_X$ or simply by $\mathbb{1}$ when X is understood, the identity endomorphism of an object X (set, vector space, algebra, etc.).

By $C_1 \sqcup C_2$ we denote the union of *disjoint* sets C_1 and C_2. Notice that this operation is strictly associative and symmetric, i.e.

$$C_1 \sqcup C_2 = C_2 \sqcup C_1 \text{ and } (C_1 \sqcup C_2) \sqcup C_3 = C_1 \sqcup (C_2 \sqcup C_3)$$

for each mutually disjoint sets C_1, C_2, and C_3.

If not specified otherwise, by a *grading* we mean a $\mathbb{Z}$ grading. The degree of a graded object will be denoted by $|w|$ though we will sometimes omit the vertical bars and write, e.g., $(-1)^{a+b}$ instead of $(-1)^{|a|+|b|}$ to save the space. We will use the Koszul sign convention meaning that whenever we commute two "things" of degrees p and q, respectively, we multiply the sign by $(-1)^{pq}$.

By Σ_n we denote, for $n \geq 1$, the *symmetric group* of n elements realized, when necessary, as the group of automorphism of the set $\{1, \ldots, n\}$. The multiplication is given by the composition of automorphisms, i.e. $\sigma\tau := \sigma \circ \tau$, and the unit is the identity. For graded indeterminates $x_1, \ldots, x_n$ and a permutation $\sigma \in \Sigma_n$ define the *Koszul sign* $\epsilon(\sigma) = \epsilon(\sigma; x_1, \ldots, x_n)$ by

$$x_1 \cdots x_n = \epsilon(\sigma; x_1, \ldots, x_n) \cdot x_{\sigma(1)} \cdots x_{\sigma(n)}, \tag{1}$$

which has to be satisfied in the free graded commutative algebra $\mathbb{k}[x_1, \ldots, x_n]$.

For graded vector spaces V, W we denote by $Lin^k(V, W)$ the vector space of degree k morphisms $V \to W$ and by $Lin(V, W)$ the graded vector space

$$Lin(V, W) := \bigoplus_{k \in \mathbb{Z}} Lin^k(V, W).$$

If W is the ground field $\Bbbk$, we obtain the *graded dual* $V^\# := Lin(V, \Bbbk)$ of V.[1] Notice that the degree k component of $V^\#$ equals the standard linear dual $(V^{-k})^\#$ of the degree $-k$ component of V. A degree k morphism $f : V \to W$ defines a map $f^\# : W^\# \to V^\#$ of the same degree by the formula

$$f^\#(\varphi) := (-1)^{k|\varphi|}\varphi \circ f. \tag{2}$$

A *dg-vector space* (abbreviating differential graded) is a couple (V, d) of a graded vector space V with a degree $+1$ differential d. The graded dual $V^\#$ of a dg-vector space is a dg-vector space, too, with the differential $d^\#$ which is the linear dual of d.

For dg-vector spaces (V_1, d_1), (V_2, d_2) we define the differential on the tensor product $V_1 \otimes V_2$ by the formula

$$d(v_1 \otimes v_2) := d_1(v_1) \otimes v_2 + (-1)^{|v_1|} v_1 \otimes d_2(v_2).$$

This, together with the flip (commutativity constrain)

$$\tau : V_1 \otimes V_2 \to V_1 \otimes V_2,\ \tau(v_1 \otimes v_2) := (-1)^{|v_1||v_2|}(v_2 \otimes v_1), \tag{3}$$

equips the category `dgVec` of dg-vector spaces and their morphisms of arbitrary degrees with a structure of symmetric monoidal category enriched over the category `Chain` of dg-vector spaces and their morphisms of degree 0 [1]. We will use a Sweedler-type notation to denote elements of the tensor product $V \otimes V$, i.e. $s \in V \otimes V$ will be written as $\sum s' \otimes s''$ or sometimes even without the summation symbol as $s' \otimes s''$.

Suppose one has a structure—algebra, module, operad, etc.—with linear operations of the form

$$\alpha : V_1' \otimes \cdots \otimes V_s' \longrightarrow V_1'' \otimes \cdots \otimes V_t'',$$

where $V_1', \ldots, V_s',\ V_1'', \ldots, V_t''$ are graded vector spaces. The dg-version of this structure—dg-algebra, dg-module, dg-operad, etc.—is the structure with the same operations, but now we assume that the graded spaces $V_1', \ldots, V_s',\ V_1'', \ldots, V_t''$ have

[1] We use # instead of the more usual $*$ to avoid confusion with the $*$ indicating a grading.

differentials and the operations satisfy

$$d''\alpha = (-1)^{|\alpha|}\alpha\, d',$$

where d' (resp. d'') is the induced differential on the product $V_1' \otimes \cdots \otimes V_s'$ (resp. on $V_1'' \otimes \cdots \otimes V_t''$).

For a graded vector space $V = \bigoplus_p V_p$ let $\uparrow V$ (resp. $\downarrow V$) denote the *suspension* (resp. the *desuspension*) of V, i.e. the graded vector space defined by $(\uparrow V)_p = V_{p-1}$ (resp. $(\downarrow V)_p = V_{p+1}$). We have the obvious natural maps $\uparrow : V \to \uparrow V$ and $\downarrow : V \to \downarrow V$.

Notation for Categories

$\texttt{Set}$	The cartesian monoidal category of sets
$\texttt{Cor}$	The category of sets finite sets and their isomorphisms
$\underline{\texttt{Cor}}$	The category of sets finite cyclically ordered sets and their isomorphisms
$\texttt{CycOp}$	The category of cyclic operads
$\texttt{CycMod}$	The category of cyclic modules
$\underline{\texttt{CycOp}}$	The category of non-Σ cyclic operads
$\underline{\texttt{CycMod}}$	The category of non-Σ cyclic modules
$\texttt{ModOp}$	The category of modular operads
$\texttt{ModMod}$	The category of modular modules
$\texttt{dgVec}$	The category of dg-vector spaces
$\texttt{Chain}$	The category of dg-vector spaces and morphisms of degree 0
$\texttt{Grp}$	The category of graphs
$\texttt{Tre}$	The category of trees

Notation for Functors

$Des : \texttt{CycOp} \to \underline{\texttt{CycOp}}$	The desymmetrization
$Sym : \underline{\texttt{CycOp}} \to \texttt{CycOp}$	The symmetrization
$\square : \texttt{CycOp} \longrightarrow \texttt{CycMod}$	The forgetful functor
$\mathrm{Mod} : \texttt{CycMod} \longrightarrow \texttt{CycOp}$	The free cyclic operad functor
$\square : \texttt{ModOp} \longrightarrow \texttt{ModMod}$	The forgetful functor
$\mathrm{Mod} : \texttt{ModMod} \longrightarrow \texttt{ModOp}$	The free modular operad functor
$\square : \texttt{ModOp} \longrightarrow \texttt{CycOp}$	The forgetful functor
$\mathrm{Mod} : \texttt{CycOp} \longrightarrow \texttt{ModOp}$	The modular completion functor

Reference

1. Mac Lane, S.: Categories for the Working Mathematician. Graduate Texts in Mathematics, vol. 5, 2nd edn. Springer, New York (1998)

Operads

6

In this chapter we recall various versions of operads required in this book. The standard references are [9] or [12], plus the original sources [2, 3] and [4].

6.1 Cyclic Operads

Consider the cobweb in Fig. 6.1 consisting of white blobs symbolizing correlation functions, and propagators represented by edges connecting some outputs of the blobs. We want to understand which abstract properties of the contractions along the propagators guarantee that the result of multiple contractions would not depend on the order in which the contractions are performed.

To be more specific, assume that the inputs of the blobs are labeled by elements of some finite sets, as $\{b, u, c, v\}$ in case of the blob y of the figure, or by $\{0, 1, 2, 3, 4\}$ in the case of the blob w. We will denote, e.g., by $x\ {}_a\circ_b\ y$ the result of the contraction along the edge e connecting the input of x labeled by a with the input of y labeled by b, see Fig. 6.1 again.

As the first step of abstraction, we want to interpret the blobs as elements of some abstract dg-vector spaces, for instance,

$$x \in \mathscr{P}\big(\{p, q, r, a\}\big),\ \ y \in \mathscr{P}\big(\{b, c, u, v\}\big),\ \text{etc.}$$

The contraction $x\ {}_a\circ_b\ y$ along the edge e is the value of a morphism

$${}_a\circ_b : \mathscr{P}\big(\{p, q, r, a\}\big) \otimes \mathscr{P}\big(\{b, c, u, v\}\big) \longrightarrow \mathscr{P}\big(\{p, q, r, c, u, v\}\big),$$

or, with the indexing sets conveniently decomposed,

$${}_a\circ_b : \mathscr{P}\big(\{p, q, r\} \cup \{a\}\big) \otimes \mathscr{P}\big(\{c, u, v\} \cup \{b\}\big) \longrightarrow \mathscr{P}\big(\{p, q, r, c, u, v\}\big).$$

© Springer Nature Switzerland AG 2020
M. Doubek et al., *Algebraic Structure of String Field Theory*, Lecture Notes in Physics 973, https://doi.org/10.1007/978-3-030-53056-3_6

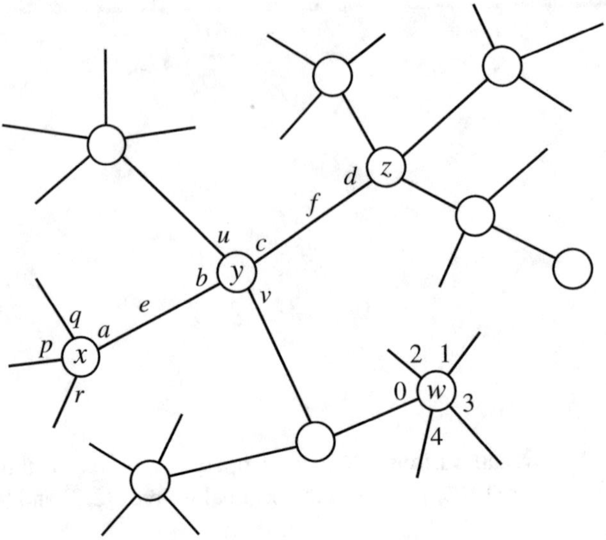

Fig. 6.1 A tree-like diagram of correlation functions and propagators

We also need a rule that would identify two blobs that differ only by relabeling the inputs. This is abstractly expressed by requiring an action on, e.g., $\mathscr{P}(\{p, q, r, a\})$ by the group of permutations of the set $\{p, q, r, a\}$.

Cyclic operads are abstractions of structures of blobs and propagators for the cases when the related diagrams are *simply connected*, i.e., when they do not have loops. The general case is covered by the notion of modular operads discussed in Sect. 6.4.

Let us proceed to a precise definition. Denote by Cor the category of finite sets and their isomorphisms. Finite sets in Cor will serve as indexing sets for the inputs of abstract blobs. In this context we call Cor the *category of corollas*, whence the notation. As usual, we denote by $A_1 \sqcup A_2$ the union of disjoint (finite) sets A_1 and A_2. Two morphisms $\rho : A_1 \to B_1$, $\sigma : A_2 \to B_2$ give the obvious induced map

$$\rho \sqcup \sigma : A_1 \sqcup A_2 \to B_1 \sqcup B_2.$$

With this terminology, we may formulate

Definition 6.1 A *cyclic operad* $\mathscr{P}$ is a family

$$\mathscr{P} = \{\mathscr{P}(S) \in \texttt{Chain} \mid S \in \texttt{Cor}\}$$

of dg-vector spaces together with degree 0 morphisms

$$\mathscr{P}(\rho) : \mathscr{P}(S) \to \mathscr{P}(T) \tag{6.1}$$

given for any isomorphism $\rho : S \to T$ of finite sets, together with degree 0 morphisms (compositions)

$$_a\circ_b : \mathscr{P}\big(S_1 \sqcup \{a\}\big) \otimes \mathscr{P}\big(S_2 \sqcup \{b\}\big) \to \mathscr{P}(S_1 \sqcup S_2) \tag{6.2}$$

defined for arbitrary disjoint finite sets S_1, S_2 and symbols a, b. These data are required to satisfy the following axioms.

(i) One has $\mathscr{P}(\mathbb{1}_S) = \mathbb{1}_{\mathscr{P}(S)}$ for any finite set S, and $\mathscr{P}(\rho\sigma) = \mathscr{P}(\rho)\mathscr{P}(\sigma)$ for arbitrary composable morphisms ρ, σ in `Cor`.

(ii) For arbitrary morphisms $\rho : S_1 \sqcup \{a\} \to T_1$ and $\sigma : S_2 \sqcup \{b\} \to T_2$ in `Cor`, one has the equality

$$\mathscr{P}\big(\rho|_{S_1} \sqcup \sigma|_{S_2}\big)\ {_a\circ_b} = {_{\rho(a)}\circ_{\sigma(b)}}\ \big(\mathscr{P}(\rho) \otimes \mathscr{P}(\sigma)\big)$$

of maps $\mathscr{P}(S_1 \sqcup \{a\}) \otimes \mathscr{P}(S_2 \sqcup \{b\}) \to \mathscr{P}\big(T_1 \sqcup T_2 \setminus \{\rho(a), \rho(b)\}\big)$.

(iii) Let $\tau : \mathscr{P}\big(S_1 \sqcup \{a\}\big) \otimes \mathscr{P}\big(S_2 \sqcup \{b\}\big) \to \mathscr{P}\big(S_2 \sqcup \{b\}\big) \otimes \mathscr{P}\big(S_1 \sqcup \{a\}\big)$ be the flip (3). One then has the equality

$$_a\circ_b = {_b\circ_a}\ \tau$$

of maps $\mathscr{P}\big(S_1 \sqcup \{a\}\big) \otimes \mathscr{P}\big(S_2 \sqcup \{b\}\big) \to \mathscr{P}(S_2 \sqcup S_1)$.[1]

(iv) For disjoint sets S_1, S_2, S_3 and symbols a, b, c, d one has the equality

$$_a\circ_b(\mathbb{1} \otimes {_c\circ_d}) = {_c\circ_d}({_a\circ_b} \otimes \mathbb{1}) \tag{6.3}$$

of maps $\mathscr{P}\big(S_1 \sqcup \{a\}\big) \otimes \mathscr{P}\big(S_2 \sqcup \{b, c\}\big) \otimes \mathscr{P}\big(S_3 \sqcup \{d\}\big) \to \mathscr{P}(S_1 \sqcup S_2 \sqcup S_3)$.

The ambient category in which cyclic operads of Definition 6.1 live is the category `Chain` of dg-vector spaces. This means that each $\mathscr{P}(S)$ is a graded vector space with a degree $+1$ differential $d = d_{\mathscr{P}}$ and that all structure operations of $\mathscr{P}$ commute with the differentials. In particular, (6.2) is a map of dg-vector spaces, where the tensor product $\mathscr{P}\big(S_1 \sqcup \{a\}\big) \otimes \mathscr{P}\big(S_2 \sqcup \{b\}\big)$ bears the differential induced in the standard manner. In elements this means that for $x \in \mathscr{P}\big(S_1 \sqcup \{a\}\big)$ and $y \in \mathscr{P}\big(S_2 \sqcup \{b\}\big)$,

$$d(x\ {_a\circ_b}\ y) = (dx)\ {_a\circ_b}\ y + (-1)^{|x|} x\ {_a\circ_b}(dy). \tag{6.4}$$

To emphasize that $\mathscr{P}$ carries a differential, we will call such a $\mathscr{P}$ sometimes more specifically a cyclic *dg*-operad. If we want to distinguish the differential $d_{\mathscr{P}}$ from other differentials that may occur in the same context, we will call it the *internal* differential.

[1]Since $S_2 \sqcup S_1 = S_1 \sqcup S_2$, $\mathscr{P}(S_2 \sqcup S_1) = \mathscr{P}(S_1 \sqcup S_2)$.

Definition 6.2 A *morphism* $\Phi : \mathscr{P} \to \mathscr{Q}$ of cyclic operads is a collection

$$\Phi = \{\Phi_S : \mathscr{P}(S) \to \mathscr{Q}(S) \mid S \in \texttt{Cor}\}$$

of degree 0 morphisms of dg-vector spaces that commute with all structure operations. This means that for $\rho : S \to T$ as in (6.1) the diagram

$$\begin{array}{ccc} \mathscr{P}(S) & \xrightarrow{\Phi_S} & \mathscr{Q}(S) \\ {\scriptstyle \mathscr{P}(\rho)}\downarrow & & \downarrow{\scriptstyle \mathscr{Q}(\rho)} \\ \mathscr{P}(T) & \xrightarrow{\Phi_T} & \mathscr{Q}(T) \end{array}$$

commutes as does, for S_1, S_2, a, b as in (6.2), the diagram

$$\begin{array}{ccc} \mathscr{P}\big(S_1 \sqcup \{a\}\big) \otimes \mathscr{P}\big(S_2 \sqcup \{b\}\big) & \xrightarrow{\ {}_a\circ_b\ } & \mathscr{P}(S_1 \sqcup S_2) \\ {\scriptstyle \Phi_{1\sqcup\{a\}} \otimes \Phi_{S_2\sqcup\{b\}}}\downarrow & & \downarrow{\scriptstyle \Phi_{S_1 \sqcup S_2}} \\ \mathscr{Q}\big(S_1 \sqcup \{a\}\big) \otimes \mathscr{Q}\big(S_2 \sqcup \{b\}\big) & \xrightarrow{\ {}_a\circ_b\ } & \mathscr{Q}(S_1 \sqcup S_2)\,. \end{array}$$

We denote by $\texttt{CycOp}$ the category of cyclic operads and their morphisms.

Remark 6.1 It should be clear that $\mathscr{P}(S)$ is an abstraction of the space of blobs with inputs indexed by the elements of the finite set S. Axiom (i) of Definition 6.1 says that the rule $S \mapsto \mathscr{P}(S)$, $\rho \mapsto \mathscr{P}(\rho)$ defines a covariant functor from the category $\texttt{Cor}$ to the category $\texttt{Chain}$ of dg-vector spaces and their degree 0 morphisms. Axiom (ii) describes the behavior of contractions with respect to reindexations.

Axiom (iii) says that, for any $x \in \mathscr{P}\big(S_1 \sqcup \{a\}\big)$ and $y \in \mathscr{P}\big(S_2 \sqcup \{b\}\big)$,

$${}_a\circ_b(x \otimes y) = (-1)^{|x||y|}\ {}_b\circ_a(y \otimes x).$$

If we write $x\ {}_a\circ_b\ y$ instead of ${}_a\circ_b(x \otimes y)$ and similarly for ${}_b\circ_a(y \otimes x)$, we see that (iii) is the graded commutativity

$$x\ {}_a\circ_b\ y = (-1)^{|x||y|}\ y\ {}_b\circ_a\ x$$

of the contractions. Informally this means that the results of the contractions of the edges in Fig. 6.1 do not depend, modulo the Koszul sign, on their orientations. Likewise, axiom (iv) requiring that

$${}_a\circ_b(\mathbb{1} \otimes {}_c\circ_d)(x \otimes y \otimes z) = {}_c\circ_d({}_a\circ_b \otimes \mathbb{1})(x \otimes y \otimes z)$$

for $x \in \mathscr{P}(S_1 \sqcup \{a\})$, $y \in \mathscr{P}(S_2 \sqcup \{b, c\})$ and $z \in \mathscr{P}(S_3 \sqcup \{d\})$, can be written as the associativity

$$x \,{}_a\circ_b (y \,{}_c\circ_d\, z) = (x \,{}_a\circ_b\, y) \,{}_c\circ_d\, z \tag{6.5}$$

of the contraction. Geometrically it means that, e.g., in Fig. 6.1, the result of the contraction would not depend on whether we contract the edge e first and then f, or vice versa.

Remark 6.2 We assume that $S \mapsto \mathscr{P}(S)$, $\rho \mapsto \mathscr{P}(\rho)$ is a *covariant* functor, so we have the *left* actions in (6.9) of the skeletal version below. The conventions when the assignment $S \mapsto \mathscr{P}(S)$ is *contravariant* and, therefore, (6.9) the *right* actions are also sometimes used in the literature. The translation between these conventions is straightforward though very technical.

Remark 6.3 The assumption that the sets S_1 and S_2 in (6.2) are disjoint is too restrictive for some applications. The remedy is to use coproducts of sets instead of their disjoint unions.

Recall that a *coproduct* of A_1 and A_2 is a set $A_1 \bigsqcup A_2$ equipped with two injections (*coprojections*) $\iota_i : A_i \to A_1 \bigsqcup A_2$, $i = 1, 2$. It is characterized by the property that, for arbitrary set S, the assignment $f \mapsto (f\iota_1, f\iota_2)$ is a one-to-one correspondence between maps $A_1 \bigsqcup A_2 \to S$ and couples (f_1, f_2) of maps $f_i : A_i \to S$. The coproduct, defined up to a canonical isomorphism, can be realized as the (ordinary) union of disjoint copies of A_1 and A_2.

With these preliminaries, we define the extended ${}_a\circ_b$-operations

$${}_a\circ_b : \mathscr{P}(S_1 \sqcup \{a\}) \otimes \mathscr{P}(S_2 \sqcup \{b\}) \to \mathscr{P}(S_1 \textstyle\bigsqcup S_2) \tag{6.6}$$

as the composition

$$\mathscr{P}(S_1 \sqcup \{a\}) \otimes \mathscr{P}(S_2 \sqcup \{b\}) \xrightarrow{\cong} \mathscr{P}(\iota_1(S_1 \sqcup \{a\})) \otimes \mathscr{P}(\iota_2(S_2 \sqcup \{b\}))$$
$$\xrightarrow{\iota_2(a) \circ_{\iota_2(b)}} \mathscr{P}(\iota_1(S_1) \sqcup \iota_2(S_2)) \xrightarrow{\cong} \mathscr{P}(S_1 \textstyle\bigsqcup S_2),$$

where ι_1, ι_2 are the coprojections for the coproduct $(S_1 \sqcup \{a\}) \bigsqcup (S_2 \sqcup \{b\})$, the first isomorphism is induced by the isomorphisms

$$S_1 \sqcup \{a\} \xrightarrow{\cong} \iota_1(S_1 \sqcup \{(a)\}) \text{ and } S_2 \sqcup \{a\} \xrightarrow{\cong} \iota_2(S_2 \sqcup \{(b)\}),$$

using the action (6.1), and the last isomorphism by the isomorphism

$$\iota_1(S_1) \sqcup \iota_2(S_2) \xrightarrow{\cong} S_1 \textstyle\bigsqcup S_2$$

using (6.1) again. Notice that $\mathscr{P}(S_1 \coprod S_2)$ as well as the map ${}_a\circ_b$ in (6.6) are defined only up to functorial canonical isomorphisms.

The composition operations in Definition 6.1 were maps

$$ {}_a\circ_b : \mathscr{P}\big(S_1 \sqcup \{a\}\big) \otimes \mathscr{P}\big(S_2 \sqcup \{b\}\big) \to \mathscr{P}(S_1 \sqcup S_2). $$

It is sometimes convenient to consider an equivalent family of compositions, namely

$$ {}_a\circ_b : \mathscr{P}(D_1) \otimes \mathscr{P}(D_2) \to \mathscr{P}\big(D_1 \sqcup D_2 \setminus \{a, b\}\big) $$

with $D_1 := S_1 \sqcup \{a\}$ and $D_2 := S_2 \sqcup \{b\}$. One can easily verify that Definition 6.1 is equivalent to

Definition 6.3 A *cyclic operad* $\mathscr{P}$ is a family

$$ \mathscr{P} = \big\{\mathscr{P}(S) \mid S \in \mathtt{Cor}\big\} $$

of dg-vector spaces together with degree 0 morphisms

$$ \mathscr{P}(\rho) : \mathscr{P}(S) \to \mathscr{P}(D) $$

given for any isomorphism $\rho : S \to D$ of finite sets, and degree 0 morphisms (compositions)

$$ {}_a\circ_b : \mathscr{P}(S_1) \otimes \mathscr{P}(S_2) \to \mathscr{P}\big(S_1 \sqcup S_2 \setminus \{a, b\}\big) $$

defined for arbitrary disjoint finite sets S_1, S_2 with elements $a \in S_1$, $b \in S_2$. These data are required to satisfy the following axioms.

(i) One has $\mathscr{P}(\mathbb{1}_S) = \mathbb{1}_{\mathscr{P}(S)}$ for any finite set S, and $\mathscr{P}(\rho\sigma) = \mathscr{P}(\rho)\mathscr{P}(\sigma)$ for arbitrary composable morphisms ρ, σ in $\mathtt{Cor}$.
(ii) For arbitrary isomorphisms $\rho : S_1 \to T_1$ and $\sigma : S_2 \to T_2$ of finite sets, one has the equality

$$ \mathscr{P}\big(\rho|_{S_1\setminus\{a\}} \sqcup \sigma|_{S_2\setminus\{b\}}\big)\ {}_a\circ_b = {}_{\rho(a)}\circ_{\sigma(b)} \big(\mathscr{P}(\rho) \otimes \mathscr{P}(\sigma)\big) $$

of maps $\mathscr{P}(S_1) \otimes \mathscr{P}(S_2) \to \mathscr{P}\big(T_1 \sqcup T_2 \setminus \{\rho(a), \sigma(b)\}\big)$.
(iii) Let $\tau : \mathscr{P}(S_1) \otimes \mathscr{P}(S_2) \to \mathscr{P}(S_2) \otimes \mathscr{P}(S_1)$ be the commutativity constraint in (3). One requires the equality

$$ {}_a\circ_b = {}_b\circ_a\, \tau $$

of maps $\mathscr{P}(S_1) \otimes \mathscr{P}(S_2) \to \mathscr{P}\big(S_2 \sqcup S_1 \setminus \{a, b\}\big)$.

(iv) For disjoint finite sets S_1, S_2, S_3 and $a \in S_1$, $b, c \in S_2$, $b \neq c$, $d \in S_3$, one has the equality

$$_a\circ_b(\mathbb{1} \otimes {}_c\circ_d) = {}_c\circ_d({}_a\circ_b \otimes \mathbb{1})$$

of maps $\mathscr{P}(S_1) \otimes \mathscr{P}(S_2) \otimes \mathscr{P}(S_3) \to \mathscr{P}\big(S_1 \sqcup S_2 \sqcup S_3 \setminus \{a, b, c, d\}\big)$.

The category Cor of finite sets is equivalent to its full skeletal subcategory whose objects are the sets $[n] := \{1, \ldots, n\}$, $n \geq 0$, with $[0]$ interpreted as the empty set $\emptyset$. It is therefore not surprising that there exists a skeletal version of Definition 6.1 in which the components of cyclic operads are not indexed by arbitrary finite sets, but by the finite ordinals $[n]$, $n \geq 1$. It can be obtained as follows.

For a non-negative integer n denote $\mathscr{P}(n) := \mathscr{P}\big([n]\big)$. The definition of the skeletal versions of the $_a\circ_b$-operations (6.2) involves, for $m, n \geq 0$, $1 \leq i \leq m+1$, $1 \leq j \leq n+1$, an isomorphism

$$\kappa = \kappa_{ij} : \big([m+1] \setminus \{i\}\big) \bigsqcup \big([n+1] \setminus \{j\}\big) \xrightarrow{\cong} [m+n] \tag{6.7}$$

given as follows. For $a \in [m+1] \setminus \{i\}$ put

$$\kappa(a) := \begin{cases} a, & 1 \leq a < i, \\ a + n - 1, & i < a \leq m+1, \end{cases}$$

while for $b \in [n+1] \setminus \{j\}$,

$$\kappa(b) := \begin{cases} b - j + i + n, & 1 \leq b < j, \\ b - j + i - 1, & j < b \leq n+1. \end{cases}$$

With these conventions, define

$$_i\bar{\circ}_j : \mathscr{P}(m+1) \otimes \mathscr{P}(n+1) \to \mathscr{P}(m+n), \ 1 \leq i \leq m+1, \ 1 \leq j \leq n+1,$$

as the composition

$$\mathscr{P}(m+1) \otimes \mathscr{P}(n+1) = \mathscr{P}([m+1]) \otimes \mathscr{P}([n+1]) \xrightarrow{{}_i\circ_j} \tag{6.8}$$
$$\mathscr{P}\big(([m+1] \setminus \{i\}) \bigsqcup ([n+1] \setminus \{j\})\big) \xrightarrow{\mathscr{P}(\kappa)} \mathscr{P}([m+n]) = \mathscr{P}(m+n),$$

where $_i\circ_j$ is the extended operation (6.6) for $S_1 = [m{+}1] \setminus \{i\}$ and $S_2 = [n{+}1] \setminus \{j\}$. Notice finally that each $\mathscr{P}(n)$ bears a natural right action of the symmetric group $\Sigma_n = Aut_{[n]}$.

Remark 6.4 The idea of the isomorphism κ in (6.7) is to apply the cyclic permutation moving $j \in [n+1]$ to $1 \in [n+1]$ to the second interval, then remove the image of j and replace $i \in [m+1]$ by the result. Alternatively, one may imagine two wheels in the plane with the spikes indexed by the ordinals $[m+1]$ resp. $[n+1]$ in the cyclic order induced by the anticlockwise orientation of the plane. Join then the ith spike of the first wheel with the jth spike of the second wheel and relabel the remaining spikes anticlockwise, starting with the spike of the first wheel labeled 1 if $i \neq 1$, or with the spike of the second wheel immediately after the spike labeled j if $i = 1$, as indicated in

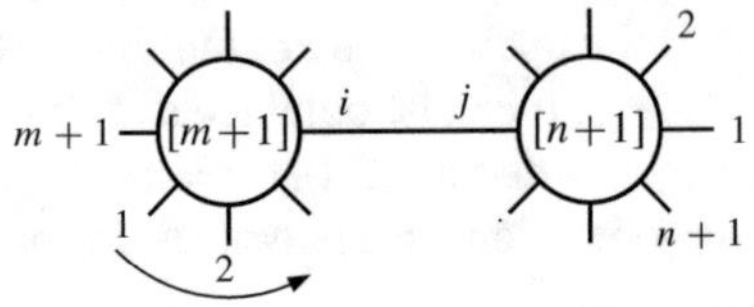

In fact, an arbitrary choice of isomorphism in (6.7) would do, but the resulting "skeletal" axioms of cyclic operads will be different.

Let us finally formulate a skeletal version of the definition of cyclic operads. We will write it in terms of elements which is in this case more convenient.

Definition 6.4 A *cyclic operad* $\mathscr{P}$ is a family

$$\mathscr{P} = \{\mathscr{P}(n) \mid n \geq 0\}$$

of dg-vector spaces together with linear left actions

$$\Sigma_n \times \mathscr{P}(n) \to \mathscr{P}(n),\ n \geq 1, \tag{6.9}$$

of the symmetric groups Σ_n, and degree 0 morphisms ('${}_i\overline{\circ}_j$-operations')

$${}_i\overline{\circ}_j : \mathscr{P}(m+1) \otimes \mathscr{P}(n+1) \to \mathscr{P}(m+n),$$

defined for arbitrary $m, n \geq 0$, $1 \leq i \leq m+1$, $1 \leq j \leq n+1$. These data are required to satisfy the following axioms.

(i) For $x \in \mathscr{P}(m+1)$, $y \in \mathscr{P}(n+1)$, $1 \leq i \leq m+1$, $1 \leq j \leq n+1$, and for permutations $\rho \in \Sigma_{n+1}$, $\sigma \in \Sigma_{m+1}$,

$$(\rho x)\ {}_{\rho(i)}\overline{\circ}_{\sigma(j)}(\sigma y) = \lambda(x\ {}_i\overline{\circ}_j\ y),$$

where $\lambda \in \Sigma_{m+n}$ is the composition

$$[m+n] \xrightarrow{\kappa_{ij}^{-1}} ([m+1]\setminus\{i\}) \sqcup ([n+1]\setminus\{j\}) \xrightarrow{\rho\sqcup\sigma}$$
$$([m+1]\setminus\{\rho(i)\}) \sqcup ([n+1]\setminus\{\sigma(j)\}) \xrightarrow{\kappa_{\rho(i)\sigma(j)}} [m+n]$$

that involves the maps (6.7).

(ii) For $x \in \mathscr{P}(m+1)$, $y \in \mathscr{P}(n+1)$, $1 \le i \le m+1$, $1 \le j \le n+1$,

$$x\ {}_i\bar{\circ}_j\ y = (-1)^{|x||y|}\lambda(y\ {}_j\bar{\circ}_i\ x),$$

where $\lambda \in \Sigma_{n+m}$ is the cyclic permutation that takes $j-i+m+1$ to 1.

(iii) For $x \in \mathscr{P}(n_1+1)$, $y \in \mathscr{P}(n_2+1)$, $x \in \mathscr{P}(n_3+1)$, $1 \le i \le n_1+1$, $1 \le k \le n_2+1$, $1 \le l \le n_3+1$, $1 \le j \le n_2+n_3$,

$$x\ {}_i\bar{\circ}_j(y\ {}_k\bar{\circ}_l\ z)$$
$$= \begin{cases} (x\ {}_i\bar{\circ}_j\ y)\ {}_{k+i-j-1}\bar{\circ}_l\ z, & j < k, \\ (-1)^{|x||y|}\lambda\cdot y\ {}_k\bar{\circ}_l(x\ {}_i\bar{\circ}_{l+j-k+1}\ z), & k \le j < k+n_3-l+1, \\ (-1)^{|x||y|}\lambda\cdot y\ {}_k\bar{\circ}_{i-j+k+n_3-1}(x\ {}_i\bar{\circ}_{l+j-k-n_3}\ z), & k+n_3-l+1 \le j < k+n_3, \\ (x\ {}_i\bar{\circ}_{j-n_3+1}\ y)\ {}_k\bar{\circ}_l\ z, & k+n_3 \le j \le n_2+n_3, \end{cases}$$

where λ is the cyclic permutation of $[n_1+n_2+n_3-1]$ taking $j+n_1-i+1$ to 1.

Remark 6.5 Let $\mathscr{P} = \{\mathscr{P}(n) \mid n \ge 0\}$ be the skeletal presentation of a cyclic operad with the structure operations ${}_i\bar{\circ}_j$ as in Definition 6.4 above. Denote by $\mathscr{Q} = \{\mathscr{Q}(n) \mid n \ge 0\}$ the collection with $\mathscr{Q}(n) := \mathscr{P}(n+1)$. Then

$$\circ_i : \mathscr{Q}(m) \otimes \mathscr{Q}(n) \to \mathscr{Q}(m+n-1),\ m, n \ge 0,$$

defined, for $1 \le i \le m$, by

$$x \circ_i y := x\ {}_i\bar{\circ}_{n+1}\ y, \tag{6.10}$$

where $x \in \mathscr{Q}(m) = \mathscr{P}(m+1)$ and $y \in \mathscr{Q}(n) = \mathscr{P}(n+1)$, are the standard $\circ_i$-operations classically used to define operads, see [7, Definition 1.1] or [12, Definition II.1.16].

Our definition of cyclic operads is slightly more general that the original one [2] in that we admit nontrivial $\mathscr{P}(S)$ in Definition 6.1 for S the empty or a one-element set, resp. nontrivial $\mathscr{P}(n)$ with $n \le 1$ in Definition 6.4. On the other hand, there is an important class of cyclic operads whose components vanish on sets of small cardinalities:

Definition 6.5 A cyclic operad $\mathscr{P}$ as in Definition 6.1 is *stable* if $\mathscr{P}(S) = 0$ for all $S \in \mathtt{Cor}$ with $\mathrm{card}(S) \leq 2$. In the skeletal setup of Definition 6.4 the stability means that $\mathscr{P}(n) = 0$ for $n \leq 2$.

The above terminology is motivated by the stability property of smooth complex projective curves of genus zero with n marked points. Such a curve is, by definition, stable, if it has no infinitesimal automorphism fixing the marked points. It is well-known that this happens if and only if $n \geq 3$. A generalization of this notion to arbitrary genera is recalled in Example 6.25.

Example 6.1 The cyclic operad $\mathscr{C}om$ is defined by

$$\mathscr{C}om(S) := \begin{cases} \Bbbk & \text{if } S \text{ has at least 3 elements, and} \\ 0 & \text{if } S \text{ has less than 3 elements.} \end{cases}$$

The functorial isomorphisms $\mathscr{C}om(\sigma) : \mathscr{C}om(S) \to \mathscr{C}om(D)$ are the identities and the compositions

$${}_a\circ_b : \mathscr{C}om\big(S_1 \sqcup \{a\}\big) \otimes \mathscr{C}om\big(S_2 \sqcup \{b\}\big) \longrightarrow \mathscr{C}om\big(S_1 \sqcup S_2\big)$$

the canonical isomorphisms $\Bbbk \otimes \Bbbk \cong \Bbbk$, $\Bbbk \otimes 0 \cong 0$, $0 \otimes \Bbbk \cong 0$ or $0 \otimes 0 \cong 0$, depending on the cardinalities of the sets S_1 resp. S_2.

For a set S with at least 3 elements denote by $\mu_S \in \mathscr{C}om(S)$ the element corresponding to $1 \in \Bbbk = \mathscr{C}om(S)$; for S with less than 3 elements we put $\mu_S := 0$. It is clear from definition that

$$\mathscr{C}om(\sigma)(\mu_S) = \mu_D$$

for any isomorphism $\sigma : S \to D$ of finite sets and that

$$\mu_{S_1 \sqcup \{a\}} \, {}_a\circ_b \, \mu_{S_2 \sqcup \{b\}} = \mu_{S_1 \sqcup S_2} \tag{6.11}$$

for arbitrary disjoint finite sets S_1 and S_2. The operad $\mathscr{C}om$ is stable. An example of a non-stable cyclic operad is provided by the endomorphism operad $\mathscr{E}nd_V$ recalled in Example 6.6 below.

It should be clear that the cyclic operad $\mathscr{C}om$ can equivalently be defined by $\mathscr{C}om(S) := \mathrm{Span}(\mu_S)$ for arbitrary finite set S, with the structure operations given by (6.11). The third kind of definition via the generating operation and a relation will be given in Proposition 6.4.

Remark 6.6 Definition 6.1 and the equivalent definitions that follow define cyclic operads in the symmetric monoidal category `Chain` of dg-vector spaces and their degree 0 morphisms. Cyclic operads can however be defined in an arbitrary symmetric monoidal category, for instance, in the cartesian monoidal category `Set` of sets. Such a `Set`-cyclic operad $\mathscr{S}$ is a collection

$$\mathscr{S} = \big\{\mathscr{S}(S) \in \mathtt{Set} \mid S \in \mathtt{Cor}\big\}$$

of sets together with maps of sets

$$\mathscr{S}(\rho) : \mathscr{S}(S) \to \mathscr{S}(D)$$

as in (6.1) and compositions

$${}_a\circ_b : \mathscr{S}\big(S_1 \sqcup \{a\}\big) \times \mathscr{S}\big(S_2 \sqcup \{b\}\big) \to \mathscr{S}(S_1 \sqcup S_2)$$

satisfying the obvious analogs of the axioms of Definition 6.1. Stability of such an operad means that $\mathscr{S}(S) = \emptyset$ if $\mathrm{card}(S) \le 2$. Each `Set`-cyclic operad $\mathscr{S}$ determines a `Chain`-operad $\mathrm{Span}(\mathscr{S})$ with

$$\mathrm{Span}(\mathscr{S})(S) := \mathrm{Span}\big(\mathscr{S}(S)\big),\ S \in \mathtt{Cor},$$

where $\mathrm{Span}\big(\mathscr{S}(S)\big)$ is the linear span of the set $\mathscr{S}(S)$. We call $\mathrm{Span}(\mathscr{S})$ the *linearization* of the `Set`-operad $\mathscr{S}$. It is clear that $\mathrm{Span}(\mathscr{S})$ is stable if and only if $\mathscr{S}$ is stable.

Example 6.2 The subcategory of stable cyclic `Set`-operads has the terminal object $*_{cyclic}$ given by

$$*_{cyclic}(S) = \begin{cases} * & \text{if } \mathrm{card}(S) \ge 3 \text{ and} \\ \emptyset & \text{otherwise,} \end{cases}$$

where $*$ is a one-point set. All its structure operations ${}_a\circ_b : * \times * \to *$ are the isomorphisms $* \times * \cong *$ and the action (6.1) is trivial. It is clear that for each stable cyclic `Set`-operad $\mathscr{S}$ there exists a unique morphism $\mathscr{S} \to *_{cyclic}$, which means that $*_{cyclic}$ is indeed a terminal `Set`-operad. The cyclic operad $\mathscr{C}om$ of Example 6.1 is the linearization of this terminal operad, that is,

$$\mathscr{C}om \cong \mathrm{Span}(*_{cyclic}). \tag{6.12}$$

Example 6.3 There exists a simple geometric interpretation of the terminal stable cyclic operad $*_{cyclic}$. For $S \in \mathtt{Cor}$, $\mathrm{card}(S) \ge 3$, denote by $M_0(S)$ the set of isomorphism classes of oriented closed surfaces of genus 0 with holes labeled by S. An example of such a surface is given in Fig. 6.2-left. For $\mathrm{card}(S) \le 2$ put $M_0(S) := \emptyset$.

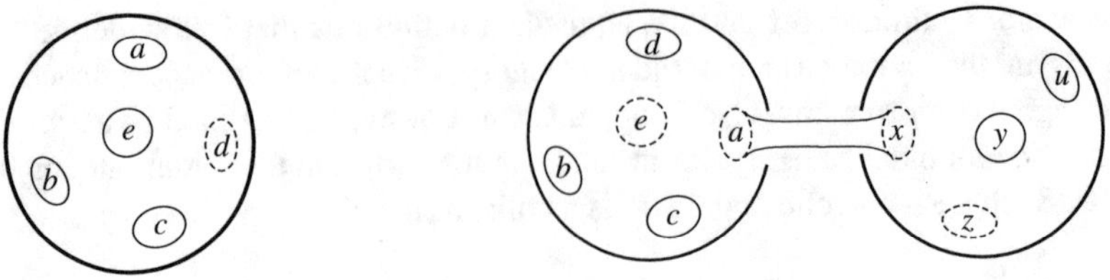

Fig. 6.2 Left: a surface representing an element in $M_0(\{a, b, c, d\})$. Right: the gluing ${}_a\circ_b$

For a surface S' with holes labeled by $S_1 \sqcup \{a\}$, and a surface S'' with holes labeled by $S_2 \sqcup \{b\}$, one has the surface $S' \,{}_a\circ_b\, S''$ obtained by connecting the circumference of the hole labeled a with the circumference of the hole labeled b using a "tube," as in Fig. 6.2-right. This operation induces a map

$${}_a\circ_b : M_0\big(S_1 \sqcup \{a\}\big) \times M_0\big(S_2 \sqcup \{b\}\big) \to M_0(S_1 \sqcup S_2)$$

of isomorphism classes which makes $M_0 = \{M_0(S)\}$ a cyclic operad. Since, for any finite set S with more than three elements there is only one isomorphism class in $M_0(S)$, one has $M_0 \cong *_{cyclic}$.

Later, in Sect. 6.3, we define algebras over cyclic operads. To do so, we will need the endomorphism operad of a dg-vector space with a non-degenerate bilinear form. This operad is generic in that all axioms of cyclic operads can be read from its properties. Before we define this operad in Example 6.7, we need to introduce a concept of multiple tensor products of graded vector spaces indexed by finite unordered sets.

Let $\{V_c\}_{c \in S}$ be such a collection of dg-vector spaces indexed by a finite set S. Since the commutativity constrain (3) is nontrivial, the multiple tensor products of V_c, $c \in S$, may depend on the order of factors. If, for instance, $S = \{a, b\}$, the space $V_a \otimes V_b$ is not the same as $V_b \otimes V_a$, only isomorphic to it via the isomorphism (3). In the presence of a grading this subtlety becomes crucial.

Since S is not a priory ordered, we want a concept that would not depend on a chosen order. The idea is to choose an order, then perform the usual tensor product, and identify the products over different orders using the Koszul sign rule. Noticing that an order of a finite set S with n elements is the same as an isomorphism $\omega : \{1, \dots, n\} \xrightarrow{\cong} S$, we are led to the following definition.

Definition 6.6 The *unordered tensor product* $\bigotimes_{c \in S} V_c$ of the collection $\{V_c\}_{c \in S}$ is the vector space of equivalence classes of usual tensor products

$$v_{\omega(1)} \otimes \cdots \otimes v_{\omega(n)} \in V_{\omega(1)} \otimes \cdots \otimes V_{\omega(n)}, \ \omega : \{1, \dots, n\} \xrightarrow{\cong} S, \tag{6.13}$$

modulo the identifications

$$v_{\omega(1)} \otimes \cdots \otimes v_{\omega(n)} \sim \epsilon(\sigma)\, v_{\omega\sigma(1)} \otimes \cdots \otimes v_{\omega\sigma(n)},\ \sigma \in \Sigma_n,$$

where $\epsilon(\sigma)$ is the Koszul sign ((1) in Part II) of the permutation σ.

Remark 6.7 The need for a subtler version of the tensor product is caused by the fact that the category dgVec of dg-vector spaces is a *non-strict* symmetric monoidal category. Similar unordered products can be defined in any symmetric monoidal category with finite colimits, see, e.g., [12, Def. II.1.58].

Let us formulate two important properties of unordered tensor products.

Lemma 6.1 *Let $\sigma : S \to D$ be an isomorphism of finite sets, $\{V_c\}_{c\in S}$ and $\{W_d\}_{d\in D}$ collections of graded vector spaces, and $\varphi = \{\varphi_c : V_c \to W_{\sigma c}\}_{c\in S}$ a family of linear maps. Then the assignment*

$$\bigotimes_{c\in S} V_c \ni \left[v_{\omega(1)} \otimes \cdots \otimes v_{\omega(n)}\right] \longmapsto \left[w_{\sigma\omega(1)} \otimes \cdots \otimes w_{\sigma\omega(n)}\right] \in \bigotimes_{d\in D} W_d$$

with $w_{\sigma\omega(i)} := \varphi_{\omega(i)}(v_{\omega(i)}) \in W_{\sigma\omega(i)}$, $1 \leq i \leq n$, defines a natural map

$$\overline{(\sigma,\varphi)} : \bigotimes_{c\in S} V_c \to \bigotimes_{d\in D} W_d$$

of unordered products

Proof. A direct verification.

A particularly important case of the above lemma is when $V_c = V_d = V$ for all $c \in S$, $d \in D$, and $\varphi_c : V \to V$ is the identity for all $c \in S$. Lemma 6.1 then gives a natural map

$$\overline{\sigma} := \overline{(\sigma,\varphi)} : \bigotimes_{c\in S} V_c \to \bigotimes_{d\in D} V_d. \tag{6.14}$$

Lemma 6.2 *For disjoint finite sets S', S'', one has a canonical isomorphism*

$$\bigotimes_{c'\in S'} V_{c'} \otimes \bigotimes_{c''\in S''} V_{c''} \cong \bigotimes_{c\in S'\sqcup S''} V_c.$$

Proof. Each $\omega' : \{1, \dots, n\} \xrightarrow{\cong} S'$ and $\omega'' : \{1, \dots, m\} \xrightarrow{\cong} S''$ determine an isomorphism

$$\omega' \sqcup \omega'' : \{1, \dots, n+m\} \xrightarrow{\cong} S' \sqcup S''$$

by the formula

$$(\omega' \sqcup \omega'')(i) := \begin{cases} \omega'(i), & \text{if } 1 \leq i \leq n, \text{ and} \\ \omega''(i-n), & \text{if } n < i \leq n+m\,. \end{cases}$$

The isomorphism of the lemma is then given by the assignment

$$[v_{\omega'(1)} \otimes \cdots \otimes v_{\omega'(n)}] \otimes [v_{\omega''(1)} \otimes \cdots \otimes v_{\omega''(m)}] \mapsto [v_{(\omega' \sqcup \omega'')(1)} \otimes \cdots \otimes v_{(\omega' \sqcup \omega'')(n+m)}].$$

Example 6.4 Let $S = \{c_1, \dots, c_n\}$. By iterating Lemma 6.2 one obtains a canonical isomorphism

$$\bigotimes_{c \in S} V_c \cong V_{c_1} \otimes \cdots \otimes V_{c_n}$$

which, crucially, depends on the order of elements of S.

Example 6.5 Let $\uparrow \Bbbk$ be the one-dimensional graded vector space concentrated in degree $+1$, $S = \{1, \dots, n\}$ and $\uparrow \Bbbk_i := \uparrow \Bbbk$ for each $i \in S$. Then $\bigotimes_{i \in S} \uparrow \Bbbk_i$ is the one-dimensional vector space concentrated in degree n. Action (6.14) applied to isomorphisms $\sigma : \{1, \dots, n\} \to \{1, \dots, n\}$, i.e., to elements of the symmetric group Σ_n, is the signum representation.

We are ready to define the endomorphism operad.

Example 6.6 Let V be a graded vector space and $s \in V \otimes V$ a symmetric degree 0 tensor. Its symmetry means that $\tau(s) = s$, where τ is the flip (3). For a finite set S put

$$\mathcal{E}nd_V(S) := Lin\big(\textstyle\bigotimes_{c \in S} V_c, \Bbbk\big) = (\textstyle\bigotimes_{c \in S} V_c)^{\#},$$

where $V_c := V$ for each $c \in S$. Given an isomorphism $\sigma : S \to D$ of finite sets, define

$$\mathcal{E}nd_V(\sigma) : \mathcal{E}nd_V(S) \to \mathcal{E}nd_V(D)$$

by $\mathcal{E}nd_V(\sigma)(f) := f\,\overline{\sigma}^{-1}$ for $f : \bigotimes_{c \in S} V_c \to \Bbbk \in \mathcal{E}nd_V(S)$ and $\overline{\sigma}$ as in (6.14).

Let S_1, S_2 be disjoint finite sets and $a \neq b$ two symbols. For $f \in \mathscr{E}nd_V(S_1 \sqcup \{a\})$ and $g \in \mathscr{E}nd_V(S_2 \sqcup \{b\})$ let $f\ {}_a\!\circ_b\ g \in \mathscr{E}nd_V(S_1 \sqcup S_2)$ be the composition

$$\bigotimes_{c\in S_1\sqcup S_2} V_c \xrightarrow{\cong} \bigotimes_{c'\in S_1} V_{c'} \otimes \bigotimes_{c''\in S_2} V_{c''} \xrightarrow{\cong} \bigotimes_{c'\in S_1} V_{c'} \otimes \Bbbk \otimes \bigotimes_{c''\in S_2} V_{c''} \tag{6.15}$$
$$\xrightarrow{\mathbb{1}\otimes s\otimes \mathbb{1}} \bigotimes_{c'\in S_1} V_{c'} \otimes V_a \otimes V_b \otimes \bigotimes_{c''\in S_2} V_{c''}$$
$$\xrightarrow{\cong} \bigotimes_{c'\in S_1\sqcup\{a\}} V_{c'} \otimes \bigotimes_{c''\in S_2\sqcup\{b\}} V_{c''} \xrightarrow{f\otimes g} \Bbbk$$

in which the isomorphisms are those of Lemma 6.2 and s is interpreted as the map $\Bbbk \to V_a \otimes V_b$ that sends $1 \in \Bbbk$ to s. Alternatively, one may define $f\ {}_a\!\circ_b\ g$ as the result of the application of the composition

$$\Big(\bigotimes_{c'\in S_1\sqcup\{a\}} V_{c'}\Big)^{\#} \otimes \Big(\bigotimes_{c''\in S_2\sqcup\{b\}} V_{c''}\Big)^{\#} \hookrightarrow \Big(\bigotimes_{c'\in S_1\sqcup\{a\}} V_{c'} \otimes \bigotimes_{c''\in S_2\sqcup\{b\}} V_{c''}\Big)^{\#} \tag{6.16}$$
$$\xrightarrow{\cong} \Big(\bigotimes_{c'\in S_1} V_{c'} \otimes V_a \otimes V_b \otimes \bigotimes_{c''\in S_2} V_{c''}\Big)^{\#} \xrightarrow{(\mathbb{1}\otimes s\otimes\mathbb{1})^{\#}} \Big(\bigotimes_{c'\in S_1} V_{c'} \otimes \Bbbk \otimes \bigotimes_{c''\in S_2} V_{c''}\Big)^{\#}$$
$$\xrightarrow{\cong} \Big(\bigotimes_{c'\in S_1} V_{c'} \otimes \bigotimes_{c''\in S_2} V_{c''}\Big)^{\#} \xrightarrow{\cong} \Big(\bigotimes_{c\in S_1\sqcup S_2} V_c\Big)^{\#}$$

to $f \otimes g \in \big(\bigotimes_{c'\in S_1\sqcup\{a\}} V_{c'}\big)^{\#} \otimes \big(\bigotimes_{c''\in S_2\sqcup\{b\}} V_{c''}\big)^{\#}$. In shorthand,

$$f\ {}_a\!\circ_b\ g = \big(\mathbb{1}_{V^{\otimes S_1}} \otimes s \otimes \mathbb{1}_{V^{\otimes S_2}}\big)^{\#}(f \otimes g)$$

and, denoting $\mathbb{1}_{S_1} := \mathbb{1}_{V^{\otimes S_1}}$ and $\mathbb{1}_{S_2} := \mathbb{1}_{V^{\otimes S_2}}$, we can write still more concisely

$$f\ {}_a\!\circ_b\ g := \big(\mathbb{1}_{S_1} \otimes s \otimes \mathbb{1}_{S_2}\big)^{\#}(f \otimes g). \tag{6.17}$$

Let us verify the associativity (6.5) of the above operations.

For $f \in \big(\bigotimes_{c'\in S_1\sqcup\{a\}} V_{c'}\big)^{\#}$, $g \in \big(\bigotimes_{c''\in S_2\sqcup\{b,c\}} V_{c'}\big)^{\#}$ and $h \in \big(\bigotimes_{c''\in S_3\sqcup\{d\}} V_{c'}\big)^{\#}$ one has

$$f\ {}_a\!\circ_b(g\ {}_c\!\circ_d\ h) = \big(\mathbb{1}_{S_1}\otimes\bar{s}\otimes\mathbb{1}_{S_2\sqcup S_3}\big)^{\#}\big(\mathbb{1}_{S_1\sqcup\{a,b\}\sqcup S_2}\otimes\bar{\bar{s}}\otimes\mathbb{1}_{S_3}\big)^{\#}(f\otimes g\otimes h), \tag{6.18}$$

while

$$(f\ {}_a\!\circ_b\ g)\ {}_c\!\circ_d\ h = \big(\mathbb{1}_{S_1\sqcup S_2} \otimes \bar{\bar{s}} \otimes \mathbb{1}_{S_3}\big)^{\#}\big(\mathbb{1}_{S_1} \otimes \bar{s} \otimes \mathbb{1}_{S_2\sqcup\{c,d\}\sqcup S_3}\big)^{\#}(f \otimes g \otimes h).$$

In the above display, $\bar{s}$ and $\bar{\bar{s}}$ are two copies of the map s. It is simple to verify that both expressions above in fact equal

$$\left(\mathbb{1}_{S_1} \otimes \bar{s} \otimes \mathbb{1}_{S_2} \otimes \bar{\bar{s}} \otimes \mathbb{1}_{S_3}\right)^{\#}(f \otimes g \otimes h)$$

which establishes (6.5). We shall however keep in mind that calculations using the shorthand (6.17) implicitly involve several canonical identifications and inclusions.

We leave as an exercise to verify that the collection $\mathcal{E}nd_V = \{\mathcal{E}nd_V(S) \mid S \in \mathtt{Cor}\}$ with the above operations fulfills also the remaining axioms of cyclic operads. The symmetry of s is necessary for axiom (iii) to hold. It is a generic example of a cyclic operad in that all the axioms can be read off from it.

Definition 6.7 The operad $\mathcal{E}nd_V$, or $\mathcal{E}nd_{(V,s)}$ if we want to stress the rôle of the symmetric tensor s, is called the *cyclic endomorphism operad* of the vector space V.

Example 6.7 We are going to describe a dual version of the endomorphism operad $\mathcal{E}nd_V$ of Example 6.6. This time V is a graded vector space equipped with a degree 0 symmetric, not necessarily non-degenerate bilinear form $B : V \otimes V \to \mathbb{k}$. For a finite set S define

$$\mathcal{D}ne_V(S) := \bigotimes\nolimits_{c \in S} V_c,$$

where $V_c := V$ for each $c \in S$. Given an isomorphism $\sigma : S \to D$ of finite sets, put

$$\mathcal{D}ne_V(\sigma) := \overline{\sigma} : \mathcal{D}ne_V(S) \to \mathcal{D}ne_V(D),$$

with $\overline{\sigma}$ as in (6.14). Let S_1, S_2 be disjoint finite sets and $a \neq b$ two symbols. Define

$${}_a\circ_b : \mathcal{D}ne_V\big(S_1 \sqcup \{a\}\big) \otimes \mathcal{D}ne_V\big(S_2 \sqcup \{b\}\big) \to \mathcal{D}ne_V\big(S_1 \sqcup S_2\big)$$

as the composition

$$\begin{aligned}\mathcal{D}ne_V\big(S_1 \sqcup \{a\}\big) \otimes \mathcal{D}ne_V\big(S_2 \sqcup \{b\}\big) &\cong \mathcal{D}ne_V\big(S_1\big) \otimes V \otimes V \otimes \mathcal{D}ne_V\big(S_2\big) \\ &\xrightarrow{\mathbb{1} \otimes B \otimes \mathbb{1}} \mathcal{D}ne_V\big(S_1\big) \otimes \mathcal{D}ne_V\big(S_2\big) \cong \mathcal{D}ne_V\big(S_1 \sqcup S_2\big)\end{aligned}$$

in which the isomorphisms are those of Lemma 6.2. In the shorthand similar to (6.17) we may write

$$x \ {}_a\circ_b \ y := (\mathbb{1}_{S_1} \otimes B \otimes \mathbb{1}_{S_2})(x \otimes y).$$

We leave again as an exercise to verify that the collection $\mathcal{D}ne_V = \{\mathcal{D}ne_V(S) \mid S \in \mathtt{Cor}\}$ with the above operations is a cyclic operad.

Let us investigate the relation between the operads $\mathscr{E}nd_V$ and $\mathscr{D}ne_V$. Assume that S is a finite set with n elements, V a graded vector space, and $V_c := V$ for each $c \in S$. We then have the *unshuffle isomorphism*

$$ush : \bigotimes_{c\in S}(V_c \otimes V_c) \xrightarrow{\cong} \bigotimes_{c\in S} V_c \otimes \bigotimes_{c\in S} V_c$$

defined, for $v'_1, \ldots, v'_n, v''_1, \ldots, v''_n \in V$, by

$$\begin{aligned} &ush\big[(v'_{\omega(1)} \otimes v''_{\omega(1)}) \otimes \cdots \otimes (v'_{\omega(n)} \otimes v''_{\omega(n)})\big] \\ &\quad := (-1)^{\varepsilon} \cdot [v'_{\omega(1)} \otimes \cdots \otimes v'_{\omega(n)}] \otimes [v''_{\omega(1)} \otimes \cdots \otimes v''_{\omega(n)}], \end{aligned}$$

where

$$\varepsilon := \sum_{1\le j<i\le n} |v'_{\omega(i)}||v''_{\omega(j)}|.$$

Each $s \in V \otimes V$ clearly determines an element

$$s^{\otimes n} := \bigotimes_{c\in S} s_c \in \bigotimes_{c\in S}(V_c \otimes V_c).$$

Using this element we define, for each finite set S, a linear map

$$\Phi_S : \mathscr{E}nd_V(S) \to \mathscr{D}ne_V(S)$$

by the formula

$$\Phi_S(f) := (f \otimes \mathbb{1})ush(s^{\otimes n}) \in \Bbbk \otimes \bigotimes_{c\in S} V_c \cong \bigotimes_{c\in S} V_c = \mathscr{D}ne_V(S), \tag{6.19}$$

for $f : \bigotimes_{c\in S} V_c \to \Bbbk \in \mathscr{E}nd_V(S)$.

Likewise, each bilinear form $B : V \otimes V \to \Bbbk$ determines a linear map

$$B^{\otimes n} := \bigotimes_{c\in S} B_c : \bigotimes_{c\in S}(V_c \otimes V_c) \to \Bbbk.$$

We define

$$\Psi_S : \mathscr{D}ne_V(S) \to \mathscr{E}nd_V(S)$$

by the formula

$$\Psi(v)(w) := B^{\otimes n}\big(ush^{-1}(v \otimes w)\big) \in \Bbbk \tag{6.20}$$

with $v, w \in \bigotimes_{c \in S} V_c$. The symmetry of s resp. of B implies that, if we take $(\mathbb{1} \otimes f)$ instead of $(f \otimes \mathbb{1})$ in (6.19) resp. $w \otimes v$ instead of $v \otimes w$ in (6.20), the resulting maps will be the same. The following lemma is easy to prove.

Lemma 6.3 *The family $\Phi = \{\Phi_S\}$ is a morphism of operads if and only if*

$$s = (\mathbb{1}_V \otimes B \otimes \mathbb{1}_V)(s \otimes s). \tag{6.21}$$

Likewise, $\Psi = \{\Psi_S\}$ is a morphism of operads if and only if

$$B = (B \otimes B)(\mathbb{1}_V \otimes s \otimes \mathbb{1}_V). \tag{6.22}$$

Remark 6.8 Equations (6.21) resp. (6.22) have simple geometric expressions. If we depict $B : V \otimes V \to \mathbb{k}$ as an abstract operation with two inputs and no output,[2] i.e.,

$B =$ 

and s as an operation with no input and two outputs, i.e.,

$s =$ ,

then (6.21) is expressed as

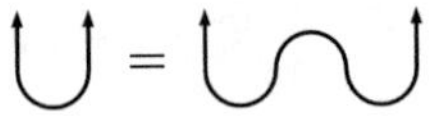

while (6.22) as

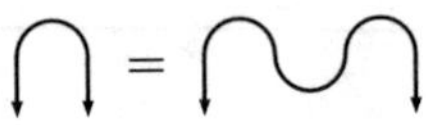

Recall that B is *non-degenerate* if, for each $x \in V$, there exists $y \in V$ such that $B(x, y) \neq 0$. It is a standard fact that this condition is equivalent to the existence of a (necessarily unique) symmetric $s \in V \otimes V$ such that

$$(\mathbb{1}_V \otimes B)(s \otimes u) = (B \otimes \mathbb{1}_V)(u \otimes s) = u, \tag{6.23}$$

for each $u \in V$ or, in pictures in the spirit of Remark 6.8,

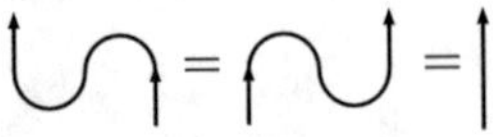

[2] Since $\mathbb{k}$ is the unit of the monoidal category dgVec, it does not count as an output.

In this situation we write $s := B^{-1}$ and call s the *Casimir element* associated with the non-degenerate bilinear form B. We say that $s \in V \otimes V$ *is non-degenerate* if there exists a non-degenerate symmetric bilinear form B such that $s = B^{-1}$.

As an exercise we recommend to verify that for B resp. s non-degenerate, (6.21) and (6.22) are automatically satisfied. The pictorial language used above makes this statement obvious. One easily proves:

Lemma 6.4 *The following conditions are equivalent.*

(i) The collection $\Phi_S : \mathscr{E}nd_V(S) \to \mathscr{D}ne_V(S)$ is an isomorphism of operads,
(ii) the collection $\Psi_S : \mathscr{D}ne_V(S) \to \mathscr{E}nd_V(S)$ is an isomorphism of operads,
(iii) the symmetric bilinear form $B : V \otimes V \to \Bbbk$ is non-degenerate and $s = B^{-1}$.

Example 6.8 We are going to describe the skeletal version of the cyclic endomorphism operad $\mathscr{E}nd_V$ introduced in Example 6.6. Notice that for $S = [n] = \{1, \ldots, n\}$, $n \geq 1$, the unordered tensor product $\bigotimes_{i \in [n]} V_i$ of Definition 6.6 is canonically isomorphic to the ordinary tensor product $V_1 \otimes \cdots \otimes V_n$. In particular, if $V_i = V$ for each $1 \leq i \leq n$, then $\bigotimes_{i \in [n]} V_i$ is canonically isomorphic to $V^{\otimes n}$, therefore

$$\mathscr{E}nd_V(n) := \mathscr{E}nd_V\big([n]\big) \cong Lin(V^{\otimes n}, \Bbbk). \tag{6.24}$$

To shorten the formulas, we will write, for e.g., $f \in \mathscr{E}nd_V(n)$, $f(v_1, \ldots, v_n)$ instead of $f(v_1 \otimes \cdots \otimes v_n)$. We will also use a variation of Sweedler's notation and write the symmetric element $s \in V \otimes V$ as the formal finite sum $s = \sum s_i' \otimes s_i''$.

Under identification (6.24), a permutation $\sigma \in \Sigma_n$ acts on a function $f \in \mathscr{E}nd_V(n)$, $n \geq 1$, by

$$(\sigma f)(v_1, \ldots, v_n) = \epsilon(\sigma) f\big(v_{\sigma(1)}, \ldots, v_{\sigma(n)}\big), \; v_1, \ldots, v_n \in V,$$

with $\epsilon(\sigma)$ the Koszul sign ((1) in Part II). For functions $f \in \mathscr{E}nd_V(m+1)$, $g \in \mathscr{E}nd_V(n+1)$, $1 \leq i \leq m+1$ and $1 \leq j \leq n+1$, one calculates that

$$\begin{aligned}(f \,{}_i\circ_j\, g)(v_1, \ldots, v_{m+n}) \\ = \sum \epsilon (-1)^{\kappa} f(v_1, \ldots, v_{i-1}, s_i', v_{i+n}, \ldots, v_{m+n}) \\ g(v_{n+i-j+1}, \ldots, v_{i+n-1}, s_i'', v_i, \ldots, v_{n+i-j})\end{aligned} \tag{6.25}$$

with

$$\kappa = |f||g| + |s_i'|(|v_{i+n}| + \cdots + |v_{m+n}|) + |s_i''|(|v_{n+i-j+1}| + \cdots + |v_{i+n-1}|) \tag{6.26}$$

and $\epsilon = \epsilon(\tau)$ the Koszul sign of the permutation

$$\tau : v_1, \ldots, v_n \longmapsto v_1, \ldots, v_{i-1}, v_{i+n}, \ldots, v_{m+n}, \\ v_{n+i-j+1}, \ldots, v_{i+n-1}, v_i, \ldots, v_{n+i-j}.$$

Let us explain how the sign in (6.25) appears. To save the space, we denote, only for the purpose of this explanation,

$$V_1 := V^{\otimes(i-1)},\ V_2 := V^{\otimes(n-j+1)},\ V_3 := V^{\otimes(j-1)},\ \text{and}\ V_4 := V^{\otimes(m-i+1)}.$$

We also denote

$$\omega_1 := v_1 \otimes \cdots \otimes v_{i-1} \in V_1,\ \omega_2 := v_i \otimes \cdots \otimes v_{n+i-j} \in V_2,$$
$$\omega_3 := v_{n+i-j+1} \otimes \cdots \otimes v_{n+i-1} \in V_3 \text{ and } \omega_4 := v_{i+n} \otimes \cdots \otimes v_{m+n} \in V_4.$$

With this notation, (6.26) reads as

$$\kappa = |f||g| + |s_i'||\omega_4| + |s_i''||\omega_3|.$$

It follows from the definition of the ${}_a\circ_b$-operations in the endomorphism operad given in Example 6.6 that the skeletal $f\ {}_i\bar{\circ}_j\ g$ is the composition of the permutation

$$\tau : V_1 \otimes V_2 \otimes V_3 \otimes V_4 \longrightarrow V_1 \otimes V_4 \otimes V_3 \otimes V_2$$

followed by the isomorphism

$$V_1 \otimes V_4 \otimes V_3 \otimes V_2 \xrightarrow{\cong} V_1 \otimes V_4 \otimes \Bbbk \otimes V_3 \otimes V_2 \tag{6.27}$$

and then by

$$(\mathbb{1}^{\otimes m} \otimes s \otimes \mathbb{1}^{\otimes n}) : V_1 \otimes V_4 \otimes \Bbbk \otimes V_3 \otimes V_2 \longrightarrow V_1 \otimes V_4 \otimes V \otimes V \otimes V_3 \otimes V_2$$

followed by the permutation

$$\rho : V_1 \otimes V_4 \otimes V \otimes V \otimes V_3 \otimes V_2 \longrightarrow V_1 \otimes V \otimes V_4 \otimes V_3 \otimes V \otimes V_2$$

and, finally, composed with

$$f \otimes g : V_1 \otimes V \otimes V_4 \otimes V_3 \otimes V \otimes V_2 \longrightarrow \Bbbk.$$

Let us inspect how the composition of the above maps acts on the element

$$v_1 \otimes \cdots \otimes v_{n+m} = \omega_1 \otimes \omega_2 \otimes \omega_3 \otimes \omega_4 \in V^{\otimes(m+n)}.$$

While

$$\tau(\omega_1 \otimes \omega_2 \otimes \omega_3 \otimes \omega_4) = \epsilon(\tau) \cdot (\omega_1 \otimes \omega_4 \otimes \omega_3 \otimes \omega_2),$$

isomorphism (6.27) brings $\omega_1 \otimes \omega_4 \otimes \omega_3 \otimes \omega_2$ into $\omega_1 \otimes \omega_4 \otimes 1 \otimes \omega_3 \otimes \omega_2$, and

$$\begin{aligned}(\mathbb{1}^{\otimes m} \otimes s \otimes \mathbb{1}^{\otimes n})&(\omega_1 \otimes \omega_4 \otimes 1 \otimes \omega_3 \otimes \omega_2) \\ &= \sum (\omega_1 \otimes \omega_4 \otimes s_i' \otimes s_i'' \otimes \omega_3 \otimes \omega_2). \end{aligned} \tag{6.28}$$

The Koszul sign rule gives

$$\begin{aligned}\sum \rho&(\omega_1 \otimes \omega_4 \otimes s_i' \otimes s_i'' \otimes \omega_3 \otimes \omega_2) \\ &= \sum (-1)^{|s_i'||\omega_4| + |s_i''||\omega_3|} (\omega_1 \otimes s_i' \otimes \omega_3 \otimes \omega_4 \otimes s_i'' \otimes \omega_2)\end{aligned}$$

and, finally,

$$\begin{aligned}(f \otimes g) &\sum (\omega_1 \otimes s_i' \otimes \omega_4 \otimes \omega_3 \otimes s_i'' \otimes \omega_2) \\ &= \sum (-1)^{(|\omega_1| + |s_i'| + |\omega_4|)|g|} f(\omega_1 \otimes s_i' \otimes \omega_4) g(\omega_3 \otimes s_i'' \otimes \omega_2) \\ &= \sum (-1)^{|f||g|} f(\omega_1 \otimes s_i' \otimes \omega_4) g(\omega_3 \otimes s_i'' \otimes \omega_2),\end{aligned}$$

where we used that $|\omega_1| + |s_i'| + |\omega_4| + |f| = 0$. The accumulated contribution of the sign factors above is precisely $\epsilon(\tau)(-1)^\kappa$ as claimed.

For $i = m+1$ and $j = 1$ (6.25) acquires a particularly nice form, namely

$$\begin{aligned}(f\ {}_{m+1}\bar{\circ}_1\ g)&(v_1, \ldots, v_{m+n}) \\ &= \sum (-1)^{|f||g|} f(v_1, \ldots, v_m, s_i') g(s_i'', v_{m+1}, \ldots, v_{m+n}).\end{aligned} \tag{6.29}$$

Example 6.9 Assume that $s \in V \otimes V$ is non-degenerate, i.e., that there exists a graded symmetric bilinear degree 0 form $B : V \otimes V \to \Bbbk$ satisfying (6.23). Then B, considered as an element of $\mathcal{E}nd_V(2)$, plays a rôle of a two-sided unit. It is indeed easy to verify that

$$f\ {}_n\bar{\circ}_1\ B = B\ {}_2\bar{\circ}_1\ f = f$$

for $f \in \mathcal{E}nd_V(n)$.

Example 6.10 The symmetric element $s = \sum s_i' \otimes s_i'' \in V \otimes V$ defines for each $n \geq 1$ a linear map

$$Lin(V^{\otimes n+1}, \Bbbk) \longrightarrow Lin(V^{\otimes n}, V),\ f \longmapsto \overline{f}, \tag{6.30}$$

by the formula

$$\tilde{f}(v_1, \dots, v_n) := \sum (-1)^{|s_i''|(|v_1|+\cdots+|v_n|)} s_i' \cdot f(v_1, \dots, v_n, s_i''),\ v_1, \dots, v_n \in V.$$

When s is the Casimir element of a non-degenerate bilinear form, the map in (6.30) is an isomorphism defining another incarnation of the skeletal endomorphism operad, namely

$$\mathcal{E}nd_V(n) \cong Lin(V^{\otimes n-1}, V),\ n \geq 1.$$

The $\circ_i$-operations in (6.10) are, for $1 \leq i \leq m$,

$$\phi \in \mathcal{E}nd_V(m+1) \cong Lin(V^{\otimes m}, V),\ \psi \in \mathcal{E}nd_V(n+1) \cong Lin(V^{\otimes n}, V)$$

and $v_1, \dots, v_{m+n-1} \in V$, given by the formula

$$\begin{aligned}&(\phi \circ_i \psi)(v_1, \dots, v_{m+n-1})\\ &= (-1)^{\kappa} \phi\big(v_1, \dots, v_{i-1}, \psi(v_i, \dots, v_{i+n-1}), v_{i+n}, \dots, v_{m+n-1}\big)\end{aligned}$$

with $\kappa = |\psi|(|v_1| + \cdots + |v_{i-1}|)$. We recognize the classical form of the $\circ_i$-operations in the endomorphism operad [9, Example 12] given by inserting ψ into the ith input of ϕ.

Example 6.11 Let us describe the skeletal version of the dual cyclic endomorphism operad $\mathcal{D}ne_V$ introduced in Example 6.7. As in (6.24) we have for each $n \geq 1$ the canonical isomorphism

$$\mathcal{D}ne_V(n) := \mathcal{D}ne_V\big([n]\big) \cong V^{\otimes n},$$

under which the symmetric group Σ_n acts by

$$\sigma(v_1 \otimes \cdots \otimes v_n) = v_{\sigma^{-1}(1)} \otimes \cdots \otimes v_{\sigma^{-1}(n)},\ v_1, \dots, v_n \in V.$$

For $v_1' \otimes \cdots \otimes v_{m+1}' \in \mathcal{D}ne_V(m+1)$ and $v_1'' \otimes \cdots \otimes v_{n+1}'' \in \mathcal{D}ne_V(n+1)$ one obtains

$$\begin{aligned}&(v_1' \otimes \cdots \otimes v_{m+1}')\, {}_i\circ_j (v_1'' \otimes \cdots \otimes v_{n+1}'')\\ &= \epsilon B(v_i', v_j'') v_1' \otimes \cdots \otimes v_{i-1}' \otimes v_{j+1}'' \otimes \cdots \otimes v_{n+1}'' \otimes v_1''\\ &\quad \otimes \cdots \otimes v_{j-1}'' \otimes v_{i+1}' \otimes \cdots \otimes v_{m+1}'\end{aligned}$$

with ϵ the Koszul sign of the permutation

$$v'_1, \ldots, v'_{m+1}, v''_1, \ldots, v''_{n+1} \longmapsto$$
$$v'_i, v''_j, v'_1, \ldots, v'_{i-1}, v''_{j+1}, \ldots, v''_{n+1}, v''_1, \ldots, v''_{j-1}, v'_{i+1}, \ldots, v'_{m+1}.$$

The "classical" $\circ_i$ operations (6.10)

$$\circ_i : \mathscr{D}ne_V(m+1) \otimes \mathscr{D}ne_V(n+1) \to \mathscr{D}ne_V(m+n),\ 1 \le i \le m,$$

are given by

$$(v'_1 \otimes \cdots \otimes v'_{m+1}) \circ_i (v''_1 \otimes \cdots \otimes v''_{n+1})$$
$$= \epsilon B(v'_i, v''_{n+1}) v'_1 \otimes \cdots \otimes v'_{i-1} \otimes v''_1 \otimes \cdots \otimes v''_n \otimes v'_{i+1} \otimes \cdots \otimes v'_m,$$

where ϵ is the Koszul sign if the permutation

$$v'_1, \ldots, v'_{m+1}, v''_1, \ldots, v''_{n+1} \longmapsto v'_i, v''_{n+1}, v'_1, \ldots, v'_{i-1}, v''_1, \ldots, v''_n, v'_{i+1}, \ldots, v'_m.$$

Let us make the intuitive concept of cyclic operads based on cobwebs presented at the beginning of this section more precise. The mathematical abstraction of a cobweb will be a graph. In the traditional approach, a graph consists of vertices and edges connecting these vertices. In the context of operads, one needs to distinguish between internal edges (those connecting two vertices) and external ones (legs) with a "free end" along which graphs can be glued together, see Definition 6.9. This needs a refinement of the classical definition. We use the one suggested by M. Kontsevich:

Definition 6.8 A *graph* Γ is a finite set $\mathit{Flag}(\Gamma)$ (whose elements are called *flags* or *half-edges*) together with an involution σ and a partition λ.

The *vertices* $\mathit{Vert}(\Gamma)$ of a graph Γ are the blocks of the partition λ. The *edges* $\mathit{Edg}(\Gamma)$ are pairs of flags forming a two-cycle of σ relative to the decomposition of a permutation into disjoint cycles. The *legs* $\mathit{Leg}(\Gamma)$ are the fixed points of σ.

We also denote by $\mathit{Leg}(v)$ the flags belonging to the block v or, in common speech, half-edges adjacent to the vertex v. The cardinality of $\mathit{Leg}(v)$ is the *valency* of v. We say that two flags $x, y \in \mathit{Flag}(\Gamma)$ *meet* if they belong to the same block of the partition λ. In plain language, this means that they share a common vertex.

One associates to a graph Γ a finite one-dimensional cell complex $|\Gamma|$, obtained by taking one copy of $[0, \frac{1}{2}]$ for each flag and imposing the following equivalence relation: The points $0 \in [0, \frac{1}{2}]$ are identified for all flags in a block of the partition λ and the points $\frac{1}{2} \in [0, \frac{1}{2}]$ are identified for pairs of flags exchanged by the involution σ. We call $|\Gamma|$ the *geometric realization* of the graph. We will sometimes make no distinction between the graph in the sense of Definition 6.8 and

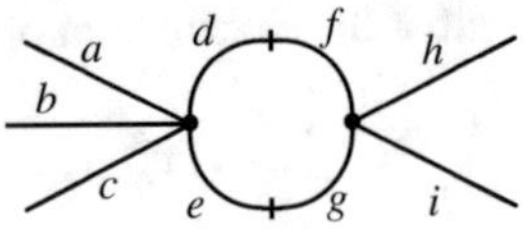

Fig. 6.3 The geometric realization of the sputnik Σ

its geometric realization. A *tree* is a graph T whose geometric realization $|T|$ is simply connected.[3]

Example 6.12 (Taken from [3]) Consider the graph Σ with $\{a, b, \ldots, i\}$ as the set of flags, the involution $\sigma = (df)(eg)$ and the partition $\{a, b, c, d, e\} \cup \{f, g, h, i\}$. Its geometric realization $|\Sigma|$ is the "sputnik" in Fig. 6.3.

Example 6.13 A S-corolla is the graph $\bigstar_S$ with one vertex and $Leg(\bigstar_S) = S$. Its geometric realization is indeed the "corolla"

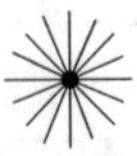

with the spikes indexed by S; whence the name and notation.

Graphs can be glued (grafted) together via their legs. Let Γ_1 be a graph with $Leg(\Gamma_1) = S_1 \sqcup \{a\}$ and Γ_2 a graph with $Leg(\Gamma_2) = S_1 \sqcup \{b\}$. We define the graph $\Gamma_1 \,{}_a\circ_b\, \Gamma_2$ by

$$Flag(\Gamma_1 \,{}_a\circ_b\, \Gamma_2) := Flag(\Gamma_1) \sqcup Flag(\Gamma_2).$$

The partition of $Flag(\Gamma_1 \,{}_a\circ_b\, \Gamma_2)$ is the union of the partitions of $Flag(\Gamma_1)$ resp. $Flag(\Gamma_2)$, and the involution σ on $Flag(\Gamma_1 \,{}_a\circ_b\, \Gamma_2)$ agrees with the involution σ_1 of $Flag(\Gamma_1)$ on $Flag(\Gamma_1) \setminus \{a\}$, with the involution σ_2 of $Flag(\Gamma_2)$ on $Flag(\Gamma_2) \setminus \{b\}$, and $\sigma(a) := b$.

Definition 6.9 We call $\Gamma_1 \,{}_a\circ_b\, \Gamma_2$ the *gluing* or *grafting* of the graphs Γ_1 and Γ_2.

In human language, $\Gamma_1 \,{}_a\circ_b\, \Gamma_2$ is obtained by gluing the free end of the leg a to the free end of b creating a new edge, symbolically:

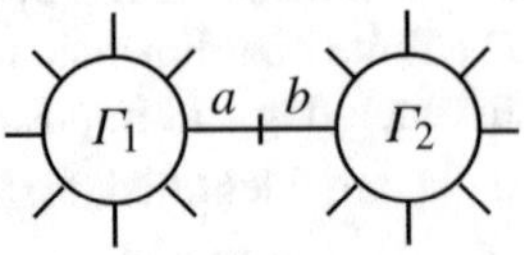

[3]Meaning that $|T|$ has no loops.

Graphs form a category. Intuitively, a *morphism* $f : \Gamma_0 \to \Gamma_1$ of graphs is given by a permutation of vertices, followed by a contraction of some edges of the graph Γ_0, leaving the legs untouched. Translated into the language of Definition 6.8, this means an injection $f^* : Flag(\Gamma_1) \to Flag(\Gamma_0)$ that commutes with the involutions. Moreover, the involution σ_0 of Γ_0 must act freely on the complement of the image of f^* in $Flag(\Gamma_0)$ (i.e., the legs of the graphs are preserved by the map f) and two flags a and b in Γ_1 meet either if $f^*(a)$ and $f^*(b)$ meet in Γ_0 or there is a chain of edges in Γ_0 from a to b.

A morphism $f : \Gamma_0 \to \Gamma_1$ clearly defines a surjective cellular map $|f| : |\Gamma_0| \to |\Gamma_1|$ of geometric realizations such that the induced map $Leg(f) : Leg(\Gamma_0) \to Leg(\Gamma_1)$ of legs is bijective. We will denote by `Grp` the category of graphs and their morphisms.

Example 6.14 There is a special class of morphisms which are given by contracting a subset of edges, without permuting the vertices. First of all, for any subset I of $Edg(\Gamma)$, there is a unique graph Γ/I such that $Flag(\Gamma/I)$ is obtained from $Flag(\Gamma)$ by deleting the flags constituting the edges in I and combining blocks of the partition that contain flags connected by a chain in I. Then the inclusion $Flag(\Gamma/I) \hookrightarrow Flag(\Gamma)$ is a morphism of graphs, which we denote by $\pi_I : \Gamma \to \Gamma/I$. An important special case is when I consists of a single edge e. We then simplify our notation by writing Γ/e instead of $\Gamma/\{e\}$ and π_e instead of $\pi_{\{e\}}$.

The graph Γ/I introduced in Example 6.14 is called the *contraction of* Γ along the set of edges I. Any morphism $f : \Gamma_0 \to \Gamma_1$ of graphs is isomorphic to a morphism of this form. This means that there exists a subset $I \subset Edg(\Gamma_0)$ and an isomorphism $\phi : \Gamma_0/I \to \Gamma_1$ such that the following diagram of graph maps commutes:

$$\begin{array}{ccccc} & & \Gamma_0/I & & \\ & \overset{\pi_I}{\nearrow} & & \overset{\phi}{\searrow} & \\ \Gamma_0 & & \xrightarrow[f]{} & & \Gamma_1 \end{array}$$

Example 6.15 The category `Grp` of graphs has two important full subcategories. The first one is the category of corollas, easily seen to be isomorphic to the category `Cor` introduced at the beginning of this section. Another important subcategory is the category `Tre` of trees consisting of graphs with simply connected geometric realizations. Notice that the only automorphism of a tree $T \in$ `Tre` fixing the legs is the identity.

Definition 6.10 A *cyclic module* is a covariant functor E : `Cor` $\to$ `Chain`. A *morphism* $\Psi : E \to F$ of cyclic modules is a natural transformation from the functor E to the functor F.

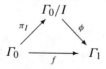

Explicitly, a cyclic module is a collection $E(S)$ of dg-vector spaces together with functorial degree 0 morphisms $E(\sigma) : E(S) \to E(T)$ specified for any isomorphism $\sigma : S \xrightarrow{\cong} T$. A morphism $\Psi : E \to F$ of cyclic modules is then a family

$$\Psi = \{\Psi_S : E(S) \to F(S) \mid S \in \texttt{Cor}\}$$

of degree 0 morphisms of dg-vector spaces such that, for each isomorphism $\rho : S \to T$ of finite sets, the diagram

$$\begin{array}{ccc} E(S) & \xrightarrow{\Psi_S} & F(S) \\ {\scriptstyle E(\rho)}\downarrow & & \downarrow{\scriptstyle F(\rho)} \\ E(T) & \xrightarrow{\Psi_T} & F(T) \end{array}$$

commutes. We denote by $\texttt{CycMod}$ the category of cyclic modules and their morphisms.

Loosely speaking, a cyclic module is a "cyclic operad without the ${}_a\circ_b$-operations." We therefore have the forgetful functor

$$\Box : \texttt{CycOp} \longrightarrow \texttt{CycMod} \tag{6.31}$$

that forgets the ${}_a\circ_b$-operations but remembers the actions of isomorphisms.

Example 6.16 Given a family $\mathfrak{S}$ of mutually non-isomorphic finite sets together with a system $G = \{G_S \mid S \in \mathfrak{S}\}$ of left *Aut*(S)-modules, there clearly exists a unique, up to isomorphism, cyclic module E_G such that, as left *Aut*(S)-modules,

$$E_G(S) = \begin{cases} G_S & \text{if } S \in \mathfrak{S}, \text{ and} \\ 0 & \text{if } S \text{ is not isomorphic to a set belonging to } \mathfrak{S}. \end{cases} \tag{6.32}$$

Such a cyclic module E_G can be constructed as follows. If S is not isomorphic to an element of $\mathfrak{S}$, we put $E(S) := 0$. In the opposite case, denote by $\overline{S} \in \mathfrak{S}$ the unique element of $\mathfrak{S}$ isomorphic to S, and choose also an isomorphism $\phi_S : S \xrightarrow{\cong} \overline{S}$. With these choices, we define $E_G(S) := E(\overline{S})$, $S \in \texttt{Cor}$, with the actions $E_G(\sigma)$ given as follows. Assume that $\rho : S' \to S''$ is an isomorphism. Then of course $\overline{S}' = \overline{S}''$ and we define the isomorphism $E(\rho) : E(S') \to E(S'')$ as the action of $\phi_{S''}\rho\phi_{S'}^{-1} \in Aut_{\overline{S}'} = Aut_{\overline{S}''}$. We call E_G the cyclic module *generated* by G.

For a graph Γ and a cyclic module E, let

$$E(\Gamma) := \bigotimes_{v \in Vert(\Gamma)} E\big(Leg(v)\big) \tag{6.33}$$

denote the unordered tensor product over the vertices of Γ. It might help to view generators of $E(\Gamma)$ as structure formulas for chemical substances represented by decorated graphs, with vertices decorated by specific atoms (elements) and internal edges of Γ representing chemical bonds.

Notice that each graph isomorphism $\phi : \Gamma_0 \xrightarrow{\cong} \Gamma_1$ induces a natural isomorphism

$$E(\phi) : E(\Gamma_0) \xrightarrow{\cong} E(\Gamma_1)$$

of dg-vector spaces. For a finite set S define

$$\mathbb{F}(E)(S) := \frac{\bigoplus_T E(T)}{\sim}, \tag{6.34}$$

where the direct sum runs over all trees T with $Leg(T) = S$, and the equivalence relation $\sim$ identifies $x \in E(T_0)$ with its image $E(\phi)(x) \in E(T_1)$ for any isomorphism $\phi : T_0 \to T_1$ such that $Leg(\phi) = \mathbb{1}_S$. For any tree T with $Leg(T) = S$ one has the canonical map

$$i_T : E(T) \to \mathbb{F}(E)(S). \tag{6.35}$$

Proposition 6.1 *The family $\mathbb{F}(E) = \{\mathbb{F}(E)(S) \mid S \in \mathtt{Cor}\}$ has a natural structure of a cyclic operad.*

Proof. Let $\rho : S \xrightarrow{\cong} D$ be an isomorphism of finite sets and T a tree with $Leg(T) = S$. There clearly exists a tree T_ρ obtained by renaming the legs of T according to ρ, together with an obvious isomorphism $\phi_\rho : T \xrightarrow{\cong} T_\rho$. The collection of induced isomorphisms

$$\big\{E(\phi_\rho) : E(T) \xrightarrow{\cong} E(T_\rho) \mid Leg(T) = S\big\}$$

clearly induces a natural isomorphism of the quotients

$$\mathbb{F}(E)(\rho) : \mathbb{F}(E)(S) \xrightarrow{\cong} \mathbb{F}(E)(D).$$

To define the ${}_a\circ_b$-operations, we recall the gluing of Definition 6.9 and notice that, by Lemma 6.2, one has for arbitrary trees T_1 and T_2, the canonical isomorphisms

$$E(T_1\,{}_a\circ_b\,T_2) \cong E(T_1) \otimes (T_2)$$

which induce morphisms (in fact, isomorphisms) of the quotients

$${}_a\circ_b : \mathbb{F}(E)\big(S_1 \sqcup \{a\}\big) \otimes \mathbb{F}(E)\big(S_2 \sqcup \{b\}\big) \to \mathbb{F}(E)\big(S_1 \sqcup S_2\big).$$

It is simple to prove that the above operations make $\mathbb{F}(E)$ a cyclic operad. It is stable if and only if $E(S) = \emptyset$ for each S with less than three elements.

Proposition 6.2 *The cyclic operad $\mathbb{F}(E)$ is free on the cyclic module E, i.e., for any cyclic operad $\mathscr{P}$ and any morphism of cyclic modules $f : E \to \mathscr{P}$[4] there exists a unique operad morphism $\Phi : \mathbb{F}(E) \to \mathscr{P}$ such that $f = \Phi \circ \iota$, where ι is the obvious inclusion $E \hookrightarrow \mathbb{F}(E)$ of cyclic modules. In diagrams:*

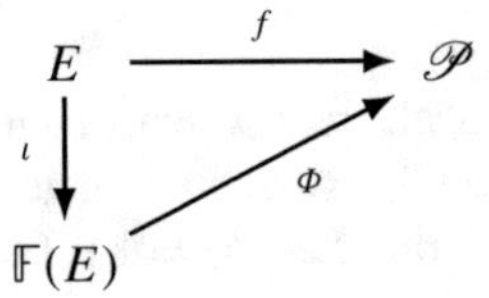

Proof. The proof follows a standard scheme, cf. [12, Proposition II.1.92], which we will not reproduce here.

The operad morphism $\Phi : \mathbb{F}(E) \to \mathscr{P}$ of Proposition 6.2 is usually called the *extension* of $f : E \to \mathscr{P}$. A concise categorical reformulation of Proposition 6.2 is that the functor

$$\mathbb{F} : \texttt{CycMod} \longrightarrow \texttt{CycOp},\ E \longmapsto \mathbb{F}(E)$$

is a *left adjoint* of the forgetful functor (6.31) which, by definition [6, p. 38], means the existence of a natural isomorphism of the morphism spaces

$$\texttt{CycMod}\big(E, \square(\mathscr{P})\big) \cong \texttt{CycOp}\big(\mathbb{F}(E), \mathscr{P}\big).$$

In the following text we will again make no notational distinction between a cyclic operad $\mathscr{P}$ and its underlying cyclic module; it will always be clear what we mean. Let

$$\Pi : \mathbb{F}(\mathscr{P}) \to \mathscr{P} \tag{6.36}$$

[4]More precisely, $f : E \to \square(\mathscr{P})$, but the implicit presence of the box is clear from the context.

be the extension of the identity map $\mathbb{1} : \mathscr{P} \to \mathscr{P}$ of cyclic modules. For a tree T with $Leg(T) = S$ we define the *contraction* along T as the composition

$$c_T : \mathscr{P}(T) \xrightarrow{i_T} \mathbb{F}(\mathscr{P})(S) \xrightarrow{\Pi} \mathscr{P}(S) \tag{6.37}$$

of Π with the canonical map (6.35).

Let us indicate an explicit construction of c_T by induction on the number of the edges of T. If T is a corolla $\bigstar_S$, $\mathscr{P}(\bigstar_S) = \mathscr{P}(S)$ and we put $c_T =: \mathbb{1}_S$. Each tree T with one edge $e = \{a, b\}$ equals $\bigstar_{S_1 \sqcup \{a\}}\, {}_a\circ_b \bigstar_{S_2 \sqcup \{b\}}$ for some finite sets S_1, S_2 with $S_1 \sqcup S_2 = S$, i.e., T looks as

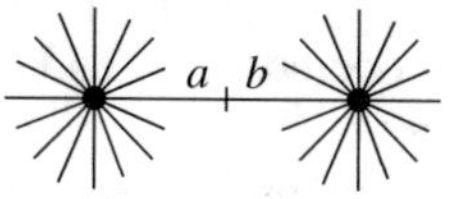

By Lemma 6.2, $\mathscr{P}(T) \cong \mathscr{P}\big(S_1 \sqcup \{a\}\big) \otimes \mathscr{P}\big(S_2 \sqcup \{b\}\big)$, and we define c_T as the structure operation ${}_a\circ_b$ of (6.2).

Suppose that T has ≥ 3 edges. Choosing one of its edges, say $e = \{a, b\}$, one decomposes $T = T_1\, {}_a\circ_b\, T_2$, where both T_1 and T_2 have less edges than T. Assume that $Leg(T_1) = S_1 \sqcup \{a\}$ and $Leg(T_2) = S_2 \sqcup \{b\}$. We then define c_T as the composition

$$\begin{aligned} c_T : \mathscr{P}(T) \cong \mathscr{P}(T_1) \otimes \mathscr{P}(T_2) &\xrightarrow{c_{T_1} \otimes c_{T_2}} \mathscr{P}\big(S_1 \sqcup \{a\}\big) \otimes \mathscr{P}\big(S_2 \sqcup \{b\}\big) \\ &\xrightarrow{{}_a\circ_b} \mathscr{P}(S_1 \sqcup S_2) = \mathscr{P}(S), \end{aligned}$$

where $c_{T_1} \otimes c_{T_2}$ has been defined by induction.

It remains to observe that the map c_T thus constructed does not depend on the choice of the edge e. Given two different edges $e = \{a, b\}$ and $f = \{c, d\}$, T looks as in

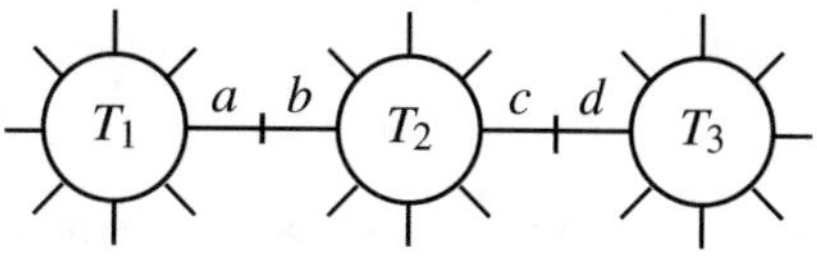

for some trees T_1, T_2, and T_3. We therefore have two ways of decomposing T,

$$T = T_1\, {}_a\circ_b (T_2\, {}_c\circ_d\, T_3) \text{ and } T = (T_1\, {}_a\circ_b\, T_2)\, {}_c\circ_d\, T_3.$$

Both cases however lead to the same result. If T_1, T_2, and T_3 are corollas, this statement is equivalent to axiom (iv) of Definition 6.1. The general case can be treated by induction.

Remark 6.9 Notice that the contraction $c_\Gamma : \mathscr{P}(\Gamma) \to \mathscr{P}(S)$ need not exist for a general graph Γ with $Leg(\Gamma) = S$. Consider, for instance, the "tick"

$\Gamma =$ 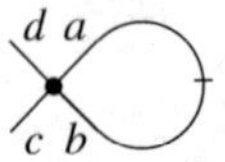

.

There is no way how to define a map

$$c_\Gamma : \mathscr{P}(\Gamma) = \mathscr{P}(\{a, b, c, d\}) \to \mathscr{P}(\{c, d\})$$

using only the structure operations of the cyclic operad $\mathscr{P}$. Operads for which such contractions exist are the modular operads recalled in Sect. 6.4.

Operads are objects of similar nature as, e.g., algebras or groups. One can therefore speak about suboperads, ideals, presentations, etc. Let us address these notions now.

Definition 6.11 A *suboperad* of $\mathscr{P}$ is a cyclic submodule $\mathscr{Q}$ of $\mathscr{P}$ closed under the structure operations of $\mathscr{P}$.

Explicitly this means that we are given, for any finite set S, a sub-dg vector space $\mathscr{Q}(S)$ of $\mathscr{P}(S)$ such that $\mathscr{P}(\rho)\big(\mathscr{Q}(S)\big) \subset \mathscr{Q}(D)$ if ρ as in (6.1) and

$$\mathscr{Q}\big(S_1 \sqcup \{a\}\big)\, {}_a\circ_b\, \mathscr{Q}\big(S_2 \sqcup \{b\}\big) \subset \mathscr{Q}(S_1 \sqcup S_2),$$

where ${}_a\circ_b$ are the structure operations of (6.2). A suboperad obviously acquires an operad structure by restricting the structure operations of $\mathscr{P}$.

Definition 6.12 An *ideal* in $\mathscr{P}$ is a cyclic submodule $\mathscr{I}$ of $\mathscr{P}$ such that

$$\mathscr{P}\big(S_1 \sqcup \{a\}\big)\, {}_a\circ_b\, \mathscr{I}\big(S_2 \sqcup \{b\}\big) \cup \mathscr{I}\big(S_1 \sqcup \{a\}\big)\, {}_a\circ_b\, \mathscr{P}\big(S_2 \sqcup \{b\}\big) \subset \mathscr{I}(S_1 \sqcup S_2)$$

for ${}_a\circ_b$ as in (6.2).

Each ideal is a suboperad but not vice versa. Important examples of ideals are (componentwise) kernels of operad morphisms. Given a morphism $\Phi : \mathscr{P} \to \mathscr{Q}$ of cyclic operads, define $\mathrm{Ker}(\Phi)$ to be the subcollection

$$\mathrm{Ker}(\Phi)(S) := \mathrm{Ker}\big(\Phi_S : \mathscr{P}(S) \to \mathscr{Q}(S)\big),\ S \in \mathit{cyclic},$$

of $\mathscr{P}$. It is simple to verify that $\mathrm{Ker}(\Phi)$ is an ideal in $\mathscr{P}$.

The characteristic property of ideals in $\mathscr{P}$ is that the operad structure of $\mathscr{P}$ induces an operad structure on the (componentwise) quotient

$$\mathscr{P}/\mathscr{I} := \{\mathscr{P}(S)/\mathscr{I}(S) \mid S \in cyclic\}.$$

Each operad is in fact a quotient of a free one, as stated in

Proposition 6.3 *Each cyclic operad $\mathscr{P}$ is isomorphic to the quotient $\mathbb{F}(E)/\mathscr{I}$ for some cyclic module E and an ideal $\mathscr{I}$ in $\mathbb{F}(E)$.*

Proof. It is clear that the map Π of (6.36) is a (componentwise) epimorphism, thus

$$\mathscr{P} \cong \mathbb{F}(\mathscr{P})/\mathrm{Ker}(\Pi). \tag{6.38}$$

As ideals, e.g., of algebras, also operadic ideals can be generated by sets of elements. Given a family $\mathfrak{S}$ of mutually non-isomorphic finite sets and a system R of elements $r_S \in \mathscr{P}(S)$, $S \in \mathfrak{S}$, one defines (R), the ideal *generated by* R, to be the smallest ideal (i.e., intersection of all ideals) in $\mathscr{P}$ containing R.

Presentation (6.38) is huge, usually much smaller ones are available. As an example we describe a small presentation of the operad $\mathscr{C}om$ recalled in Example 6.1. Consider the family $\mathfrak{S}$ consisting of a single set $\{1, 2, 3\}$ and take

$$G_{\{1,2,3\}} := \mathrm{Span}(\mu_{\{1,2,3\}}) \cong \Bbbk$$

with the trivial action of the symmetric groups $\Sigma_3 = Aut(\{1, 2, 3\})$. Let E_μ be the cyclic module generated, in the sense of (6.32), by the generating system G and (R) be the ideal in $\mathbb{F}(E_\mu)$ generated by the single element

$$r_{\{1,2,5,6\}} := \mu_{\{1,2,3\}} \, {}_3{\circ}_4 \, \mu_{\{4,5,6\}} - \mu_{\{2,5,3\}} \, {}_3{\circ}_4 \, \mu_{\{4,6,1\}} \in \mathbb{F}(E_\mu)(\{1, 2, 5, 6\})$$

graphically expressed as

$r_{\{1,2,5,6\}} =$ 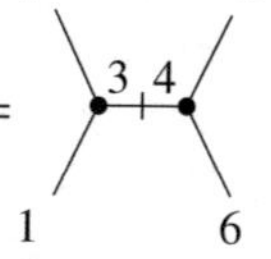 $-$ 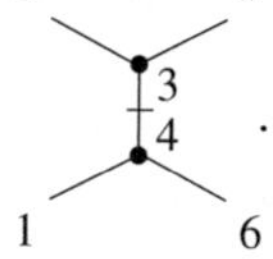 .

Proposition 6.4 *The cyclic operad $\mathscr{C}om$ has the presentation*

$$\mathscr{C}om \cong \mathbb{F}(E_\mu)/(R). \tag{6.39}$$

Proof. It is clear from definitions that $\mathbb{F}(E_\mu)(S)$ is spanned by trees with trivalent vertices and legs indexed by S. Modding out by (R) means identifying two such trees that differ by a finite sequence of the moves

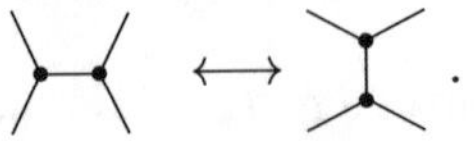

Since *arbitrary* trees generating $\mathbb{F}(E_\mu)(S)$ can clearly be related by a sequence of such moves, we see that $\mathbb{F}(E_\mu)/(R)(S)$ is one-dimensional, with the trivial actions of the automorphism group. Isomorphism (6.39) is clear now.

Proposition 6.4 can be reformulated by saying that $\mathcal{C}om$ is generated by a fully symmetric element $\mu_{\{1,2,3\}} \in \mathcal{C}om(\{1,2,3\})$ such that

$$\mu_{\{1,2,3\}}\ {}_3\!\circ_3\ \mu_{\{3,4,5\}} \tag{6.40}$$

is cyclically symmetric in $1, 2, 4, 5$.[5] Equation (6.40) in this setup requires explanation. Given $\mu_{\{1,2,3\}}$ determines the generator $\mu_S \in \mathcal{C}om(S)$ for an arbitrary S with three elements. Indeed, choose an isomorphism $\sigma : \{1,2,3\} \to S$ and put $\mu_S := \mathcal{C}om(\sigma)(\mu_{\{1,2,3\}})$. The result does not depend on the choice of σ by the symmetry of $\mu_{\{1,2,3\}}$. The element $\mu_{\{3,4,5\}}$ in (6.40) is the one determined by $\mu_{\{1,2,3\}}$ in this way.

6.2 Non-Σ Cyclic Operads

Let us mention an important variant of operads, namely *non-Σ* cyclic[6] operads obtained by taking, in Definition 6.1, instead of the category $\mathtt{Cor}$ of finite sets and their isomorphisms the category $\underline{\mathtt{Cor}}$ of finite cyclically ordered sets and isomorphisms preserving the cyclic orders.

A non-Σ cyclic operad is thus a collection $\underline{\mathcal{P}} = \{\underline{\mathcal{P}}(S) \mid S \in \underline{\mathtt{Cor}}\}$ of dg-vector spaces, but the actions (6.1) are this time defined only for order-preserving isomorphisms $\sigma : S \to D$ of cyclically ordered sets. In (6.2) we assume that the sets $S_1 \sqcup \{a\}$ res. $S_2 \sqcup \{b\}$ are cyclically ordered and that $S_1 \sqcup S_2$ has the obvious induced cyclic order. Otherwise, the axioms are formally the same as in Definition 6.1. We denote by $\underline{\mathtt{CycOp}}$ the category of non-Σ cyclic operads. As expected, a non-Σ cyclic operad $\underline{\mathcal{P}}$ is *stable* if $\underline{\mathcal{P}}(S) = 0$ whenever $\mathrm{card}(S) \le 2$.

[5]Thanks to the symmetry of $\mu_{\{1,2,3\}}$, (6.40) is actually *fully* symmetric in 1, 2, 3, 4.

[6]Instead of "non-Σ" also the prefix *non-symmetric* is sometimes used in the literature.

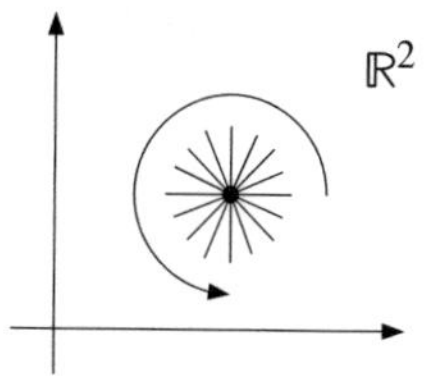

Fig. 6.4 The cyclic order on the set of half-edges of a vertex of a planar graph induced by the anticlockwise orientation of the plane

It is clear that each cyclic operad determines, by restricting to cyclically ordered sets, a non-Σ cyclic operad. This gives rise to the forgetful functor (the *desymmetrization*[7])

$$Des : \mathtt{CycOp} \longrightarrow \underline{\mathtt{CycOp}} \tag{6.41}$$

that has a left adjoint (the *symmetrization*)

$$Sym : \underline{\mathtt{CycOp}} \to \mathtt{CycOp}. \tag{6.42}$$

We leave as an exercise to describe $Sym(\underline{\mathscr{P}})$ of a non-Σ cyclic operad explicitly.

A functor $\underline{E} : \underline{\mathtt{Cor}} \to \mathtt{Chain}$ will be called a *non-Σ cyclic module*. We again have the forgetful functor

$$\square : \underline{\mathtt{CycOp}} \longrightarrow \underline{\mathtt{CycMod}} \tag{6.43}$$

from the category of non-Σ cyclic operads to the category of non-Σ cyclic modules. For such a module $\underline{E}$, one has the free non-Σ cyclic operad $\underline{\mathbb{F}}(\underline{E})$ given by a formula similar to (6.34) but involving only *planar* trees, not arbitrary ones. Let us give

Definition 6.13 A *planar graph* is a graph Γ as in Definition 6.8 together with an (isotopy class of an) embedding of its geometric realization $|\Gamma|$ into the oriented plane $\mathbb{R}^2$.

The embedding $|\Gamma| \hookrightarrow \mathbb{R}^2$ induces for each vertex v of Γ a cyclic order on the set $Leg(v)$ of half-edges adjacent to v, as indicated in Fig. 6.4. In the same manner also the set $Leg(\Gamma)$ acquires a cyclic order. The converse is true for trees:

Proposition 6.5 *An arbitrary choice of cyclic orders of the sets of half-edges adjacent to the vertices of a tree T is induced by (a unique isotopy class of) an embedding of $|T|$ into the oriented plane $\mathbb{R}^2$. Therefore, a planar structure on a tree is the same as specifying the cyclic orders of $Leg(v)$ for each vertex v of T.*

[7]Not to be mistaken with Batanin's desymmetrization of [1].

Proposition 6.4 is so obvious that we are not going to prove it here. Notice that it does not hold for general graphs. As an example, consider the graph

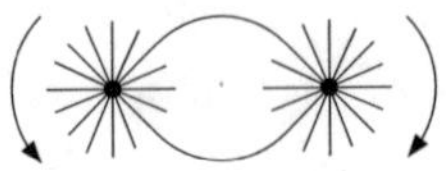

with two vertices whose half-edges are cyclically oriented as indicated by the arrows. This graph cannot be embedded into the plane such that these orientations are induced by the embedding.

An *isomorphism* of planar graphs is an isomorphism preserving the prescribed cyclic orders of the corresponding sets of half-edges. For a planar graph Γ and a non-Σ cyclic module $\underline{E}$ define

$$\underline{E}(\Gamma) := \bigotimes_{v \in Vert(\Gamma)} \underline{E}\big(Leg(v)\big).$$

The above formula makes sense since the sets $Leg(v)$ are cyclically ordered by assumption. As before, an isomorphism $\phi : \Gamma_0 \xrightarrow{\cong} \Gamma_1$ of planar graphs induces an isomorphism

$$\underline{E}(\phi) : \underline{E}(\Gamma_0) \xrightarrow{\cong} \underline{E}(\Gamma_1).$$

For a finite cyclically ordered set $S \in \underline{\mathtt{Cor}}$ define

$$\underline{\mathbb{F}}(\underline{E})(S) := \frac{\bigoplus_T \underline{E}(T)}{\sim},$$

where the direct sum this time, unlike in (6.34), runs over *planar* trees T with $Leg(T) = S$ (equality of cyclically ordered sets); the equivalence relation is an obvious analog of that in (6.34). One has the expected

Proposition 6.6 *The family* $\underline{\mathbb{F}}(E) = \{\underline{\mathbb{F}}(\underline{E})(S) \mid S \in \underline{\mathtt{Cor}}\}$ *is a non-*Σ *cyclic operad.*

The proof is a verbatim analog of the proof of Proposition 6.2, one only needs to observe that the gluing $T_1\, {}_a\circ_b\, T_2$ of two planar trees is planar again. One finally has:

Proposition 6.7 *The non-*Σ *cyclic operad* $\underline{\mathbb{F}}(\underline{E})$ *is free on the* $\underline{\mathtt{Cyc}}$*-module* $\underline{E}$*.*

Example 6.17 For a finite cyclically ordered set $S \in \underline{\mathtt{Cor}}$, put

$$\underline{\mathcal{A}ss}(S) := \begin{cases} \Bbbk & \text{if } S \text{ has at least 3 elements, and} \\ 0 & \text{if } S \text{ has less than 3 elements.} \end{cases}$$

The action $\underline{\mathcal{A}ss}(\sigma)$ of an order-preserving isomorphisms of cyclically ordered set is the identity, and the compositions

$$_a\circ_b : \underline{\mathcal{A}ss}\big(S_1 \sqcup \{a\}\big) \otimes \underline{\mathcal{A}ss}\big(S_2 \sqcup \{b\}\big) \longrightarrow \underline{\mathcal{A}ss}\big(S_1 \sqcup S_2\big)$$

are the canonical isomorphisms. Then $\underline{\mathcal{A}ss}$ with the above structure is a stable non-Σ cyclic operad.

The definition of the operad $\underline{\mathcal{A}ss}$ looks very much the same as the definition of $\mathcal{C}om$ given in Example 6.1, but one must remember that these operads belong to different categories. The suboperads, ideals, presentations, and the related notions for non-Σ cyclic operads can be defined analogously as for ordinary operads; we leave the details to the reader. The following statement describes a presentation of the non-Σ cyclic operad $\underline{\mathcal{A}ss}$.

Proposition 6.8 *The non-Σ cyclic operad $\underline{\mathcal{A}ss}$ is generated by an element $\mu_{\{1,2,3\}} \in \underline{\mathcal{A}ss}(\{1,2,3\})$ such that*

$$\mu_{\{1,2,3\}}\ {}_3\circ_3\ \mu_{\{3,4,5\}} \tag{6.44}$$

is cyclically symmetric in $1, 2, 4, 5$.

Example 6.18 The symmetrization (6.42) $\mathcal{A}ss := Sym(\underline{\mathcal{A}ss})$ has a nice explicit description. For a set with $n \geq 3$ elements one has

$$\mathcal{A}ss(S) = \mathrm{Span}\Big(\frac{\omega : \{1, \ldots, n\} \xrightarrow{\cong} S}{\sim} \Big),$$

the vector space spanned by the set of isomorphisms ω between $\{1, \ldots, n\}$ and S, modulo the relation $\sim$ that identifies, for each *cyclic* permutation $\lambda \in \Sigma_n$, the isomorphism ω with $\omega\lambda$.

It is convenient to denote the equivalence class of $[\omega]$ by $(c_1, \ldots, c_n)$, with $c_i := \omega(i)$, $1 \leq i \leq n$. By the definition of the equivalence, one has the equality

$$(c'_1, \ldots, c'_n) = (c''_1, \ldots, c''_n) \tag{6.45}$$

if and only if $c'_i = c''_{\lambda(i)}$ for some cyclic permutation $\lambda \in \Sigma_n$, $1 \leq i \leq n$.

In this language, the action (6.1) is given by

$$\mathcal{A}ss(\sigma)(c_1, \ldots, c_n) = (\sigma c_1, \ldots, \sigma c_n).$$

Let us describe the compositions (6.2). Thanks to equality (6.45), every element of $\mathscr{A}ss\big(S_1 \sqcup \{a\}\big)$ can be (uniquely) represented as $(c'_1, \ldots, c'_n, a)$ and, likewise, each element of $\mathscr{A}ss\big(S_2 \sqcup \{b\}\big)$ can be represented as $(b, c''_1, \ldots, c''_m)$. We then have

$$(c'_1, \ldots, c'_n, a)\,{}_a\circ_b(b, c''_1, \ldots, c''_m) := (c'_1, \ldots, c'_n, c''_1, \ldots, c''_m).$$

Example 6.19 The category of stable non-Σ-cyclic operads in the category `Set` of sets has the terminal object $\underline{*}_{cyclic}$ given by

$$\underline{*}_{cyclic}(S) := \begin{cases} * & \text{if } \mathrm{card}(S) \geq 3 \text{ and} \\ \emptyset & \text{otherwise,} \end{cases}$$

where $*$ is a one-point set, with all structure operations the isomorphisms $* \times * \xrightarrow{\cong} *$. The non-$\Sigma$ cyclic operad $\underline{\mathscr{A}ss}$ of Example 6.17 is the linearization of this operad,

$$\underline{\mathscr{A}ss} \cong \mathrm{Span}(\underline{*}_{cyclic}).$$

Example 6.20 As the terminal stable cyclic operad $*_{cyclic}$ of Example 6.2 has an interpretation in terms of oriented genus zero surfaces with holes, the operad $\underline{*}_{cyclic}$ is isomorphic with the operad W_0 whose components $W_0(S)$ for $\mathrm{card}(S) \geq 3$, consist of isomorphism classes of planar cogwheels

whose teeth are indexed by the cyclically ordered set S, with the cyclic order agreeing with the one induced by the anti-clockwise orientation of the plane. If $\mathrm{card}(S) \leq 2$ we put $W_0(S) := \emptyset$.

The structure operations are induced by gluing these cogwheels together along the tips of their teeth so that the orientation is preserved:

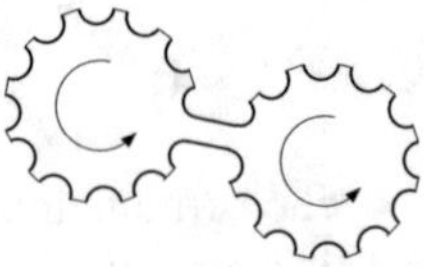

Another presentation of elements of W_0, closer to the open string philosophy, is via oriented spheres with one toothed hole:

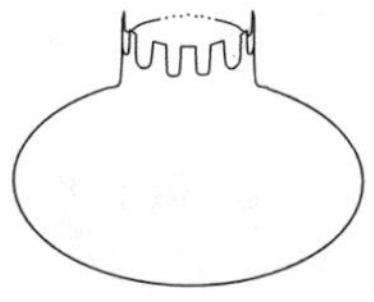

whose teeth are indexed by S so that its cyclic order agrees with the one induced by the orientation of the sphere.

6.3 Operad Algebras

The importance of operads is that they describe, via their representations, algebras of specific types.

Definition 6.14 Let V be a graded vector space and $s \in V \otimes V$ a symmetric degree 0 element. An *algebra* over a cyclic operad $\mathscr{P}$, or a $\mathscr{P}$*-algebra*, is a morphism $\alpha : \mathscr{P} \to \mathscr{E}nd_{(V,s)}$.

Example 6.21 An algebra over the operad $\mathscr{C}om$ from Example 6.1 is the same as a fully symmetric degree 0 linear map $f : V^{\otimes 3} \to \Bbbk$ such that the linear map $V^{\otimes 4} \to \Bbbk$

$$\sum f(v_1, v_2, s_i')f(s_i'', v_3, v_4) \tag{6.46}$$

is cyclically symmetric in v_1, v_2, v_3, and v_4.[8] Let us verify this statement.

A $\mathscr{C}om$-algebra is a morphism $\alpha : \mathscr{C}om \to \mathscr{E}nd_{(V,s)}$. By Proposition 6.11, such a morphism is determined by $f := \alpha(\mu_{\{1,2,3\}})$, where $\mu_{\{1,2,3\}}$ satisfies (6.40). Such an f is fully symmetric by the symmetry of $\mu_{\{1,2,3\}}$. Since α is an operad morphism,

$$\alpha(\mu_{\{1,2,3\}}\ {}_3\circ_3\ \mu_{\{3,4,5\}}) = \alpha(\mu_{\{1,2,3\}})\ {}_3\circ_3\ \alpha(\mu_{\{3,4,5\}}),$$

while it is simple to identify, invoking the definition of the composition in the endomorphism operad, the right-hand side term with expression (6.46). The cyclic symmetry of (6.40) therefore implies the cyclic symmetry of (6.46). On the other hand, it is easy to see that each $f : V^{\otimes 3} \to \Bbbk$ with the above properties determines a unique morphism $\alpha : \mathscr{C}om \to \mathscr{E}nd_{(V,s)}$ such that $f := \alpha(\mu_{\{1,2,3\}})$.

[8] As before, we are using a variation on Sweedler's convention $s = \sum s_i' \otimes s_i''$.

One can also give the following more standard incarnation of $\mathcal{C}om$-algebras. Define a degree 0 bilinear operation $\mu : V \otimes V \to V$ by

$$\mu(a, b) := \sum s'_i f(s''_i, a, b),\ a, b \in V.$$

If s is non-degenerate and $s^{-1} = B$, then the assignment $f \leftrightarrow \mu$ defines a one-to-one correspondence between $\mathcal{C}om$-algebra structures f and commutative associative multiplications μ on V such that

$$B\big(\mu(a, b), c\big) = B\big(a, \mu(b, c)\big).$$

These structures are commutative non-unital versions of *Frobenius algebras*.

Let us denote by $\underline{\mathcal{E}nd}_{(V,s)}$ the endomorphism operad $\mathcal{E}nd_{(V,s)}$ considered as a non-Σ cyclic operad, i.e.,

$$\underline{\mathcal{E}nd}_{(V,s)} := \square(\mathcal{E}nd_{(V,s)}),$$

where $\square$ is the forgetful functor of (6.41). By the standard properties of adjunctions, for any non-Σ cyclic operad $\underline{\mathcal{P}}$ there exists a one-to-one correspondence between morphism

$$\alpha : Sym(\underline{\mathcal{P}}) \to \mathcal{E}nd_{(V,s)}$$

in the category of cyclic operads, and morphism

$$\underline{\alpha} : \underline{\mathcal{P}} \to \underline{\mathcal{E}nd}_{(V,s)} \tag{6.47}$$

in the category of non-Σ cyclic operads. Therefore an algebra over the symmetrization $Sym(\underline{\mathcal{P}})$ is the same as a morphism (6.47). The description of $\mathcal{A}ss = Sym(\underline{\mathcal{A}ss})$-algebras in the following example thus easily follows from Proposition 6.8.

Example 6.22 An algebra over the operad $\mathcal{A}ss$ from Example 6.18 is the same as a degree 0 cyclically symmetric linear map $f : V^{\otimes 3} \to \Bbbk$ such that the linear map $V^{\otimes 4} \to \Bbbk$ defined by

$$\sum f(v_1, v_2, s'_i) f(s''_i, v_3, v_4)$$

is cyclically symmetric, too.

Define as in Example 6.21 $\mu : V \otimes V \to V$ by $\mu(a, b) := \sum s'_i f(s''_i, a, b)$. If s is non-degenerate and $s^{-1} = B$, then the assignment $f \mapsto \mu$ defines an one-to-one correspondence between $\mathcal{A}ss$-algebra structures f and associative, not necessarily commutative, multiplications μ on V such that

$$B\big(\mu(a,b),c\big) = B\big(a,\mu(b,c)\big).$$

So $\mathscr{A}ss$-algebras in the above sense are non-unital *Frobenius algebras* with s the corresponding *Casimir element*.

6.4 Modular Operads

While cyclic operads were abstractions of structures of blobs and propagators with simply connected underlying graphs, modular operads have arbitrary graphs as their pasting schemes. As a result, there is another grading by the genus of the underlying graph. Let us give a general definition. Let $\mathbb{A}$ be an abelian semigroup, i.e., a set with an associative commutative operation $+ : \mathbb{A} \times \mathbb{A} \to \mathbb{A}$ and a unit $0 \in \mathbb{A}$. A typical example will be the semigroup $\mathbb{N} = \{0, 1, 2, \ldots\}$ of natural numbers. When convenient, we consider $\mathbb{A}$ as a discrete category.

Definition 6.15 A *modular module* is a covariant functor

$$E : \texttt{Cor} \times \mathbb{A} \to \texttt{Chain}.$$

A *morphism* $\Psi : E \to F$ of modular modules is a natural transformation from the functor E to the functor F.

Explicitly, a modular module E is a collection $E(S; g)$, $S \in \texttt{Cor}$, $g \in \mathbb{A}$, of dg-vector spaces together with functorial degree 0 morphisms

$$E(\sigma) : E(S; g) \to E(T; g) \tag{6.48}$$

specified for any isomorphism $\sigma : S \xrightarrow{\cong} T$ and $g \in \mathbb{A}$. We call $g \in \mathbb{A}$ the *operadic genus* or simply the *genus* of the component $E(S; g)$ of E. A morphism $\Psi : E \to F$ of modular modules is then a family

$$\Psi = \{\Psi(S; g) : E(S; g) \to F(S; g) \mid (S; g) \in \texttt{Cor} \times \mathbb{A}\} \tag{6.49}$$

of degree 0 morphisms of dg-vector spaces such that, for each isomorphism $\rho : S \to T$ of finite sets, the diagram

$$\begin{array}{ccc} E(S; g) & \xrightarrow{\Psi_S} & F(S; g) \\ {\scriptstyle E(\rho)}\downarrow & & \downarrow{\scriptstyle F(\rho)} \\ E(T; g) & \xrightarrow{\Psi_T} & F(T; g) \end{array}$$

commutes. We denote by $\texttt{ModMod}$ the category of modular modules.

Remark 6.10 Later on, we will also need *degree-k* morphisms of modular modules for an integer k. It is a family as in (6.49), but this time consisting of morphisms of degree k. For $k = 0$ this extended notion of course agrees with Definition 6.15.

Definition 6.16 Let $s \in \mathbb{A}$ be a chosen element called the *step*. A *modular operad* with step s is a modular module

$$\mathcal{M} = \{\mathcal{M}(S; g) \in \mathtt{Chain} \mid (S; g) \in \mathtt{Cor} \times \mathbb{A}\} \tag{6.50}$$

together with degree 0 morphisms (compositions)

$$_a\circ_b : \mathcal{M}(S_1 \sqcup \{a\}; g_1) \otimes \mathcal{M}(S_2 \sqcup \{b\}; g_2) \to \mathcal{M}(S_1 \sqcup S_2; g_1 + g_2) \tag{6.51}$$

defined for arbitrary disjoint finite sets S_1, S_2, symbols a, b, and arbitrary genera $g_1, g_2 \in \mathbb{A}$. There are, moreover, degree 0 contractions

$$\circ_{uv} = \circ_{vu} : \mathcal{M}(S \sqcup \{u, v\}; g) \to \mathcal{M}(S; g + s) \tag{6.52}$$

given for any finite set S, genus $g \in \mathbb{A}$, and symbols u, v. These data are required to satisfy the following axioms.

(i) For arbitrary isomorphisms $\rho : S_1 \sqcup \{a\} \to T_1$ and $\sigma : S_2 \sqcup \{b\} \to T_2$ of finite sets and genera $g_1, g_2 \in \mathbb{A}$, one has the equality

$$\mathcal{M}(\rho|_{S_1} \sqcup \sigma|_{S_2})\, {_a\circ_b} = {_{\rho(a)}\circ_{\sigma(b)}} \big(\mathcal{M}(\rho) \otimes \mathcal{M}(\sigma)\big)$$

of maps

$$\mathcal{M}(S_1 \sqcup \{a\}; g_1) \otimes \mathcal{M}(S_2 \sqcup \{b\}; g_2) \to \mathcal{M}(T_1 \sqcup T_2 \setminus \{\rho(a), \sigma(b)\}; g_1 + g_2).$$

(ii) For an isomorphism $\rho : S \sqcup \{u, v\} \to T$ of finite sets and a genus $g \in \mathbb{A}$, one has the equality

$$\mathcal{M}(\rho|_S)\, \circ_{uv} = \circ_{\rho(u)\rho(v)} \mathcal{M}(\rho) \tag{6.53}$$

of maps $\mathcal{M}(S \sqcup \{u, v\}; g) \to \mathcal{M}(T \setminus \{\rho(u), \rho(v)\}; g + s)$.

(iii) For S_1, S_2, a, b and g_1, g_2 as in (6.51), one has the equality

$$_a\circ_b = {_b\circ_a}\, \tau \tag{6.54}$$

of maps $\mathcal{M}(S_1 \sqcup \{a\}; g_1) \otimes \mathcal{M}(S_2 \sqcup \{b\}; g_2) \to \mathcal{M}(S_1 \sqcup S_2; g_1 + g_2)$.[9]

[9]Recall that τ is the commutativity constraint in the category of graded vector spaces.

(iv) For mutually disjoint sets S_1, S_2, S_3, symbols a, b, c, d, and genera $g_1, g_2, g_3 \in \mathbb{A}$, one has the equality

$$ {}_a\circ_b(\mathbb{1} \otimes {}_c\circ_d) = {}_c\circ_d({}_a\circ_b \otimes \mathbb{1}) \tag{6.55}$$

of maps from $\mathcal{M}(S_1 \sqcup \{a\}; g_1) \otimes \mathcal{M}(S_2 \sqcup \{b, c\}; g_2) \otimes \mathcal{M}(S_3 \sqcup \{d\}; g_3)$ to the space $\mathcal{M}(S_1 \sqcup S_2 \sqcup S_3; g_1+g_2+g_3)$.

(v) For a finite set S, symbols a, b, c, d and a genus $g \in \mathbb{A}$ one has the equality

$$\circ_{ab} \ \circ_{cd} = \circ_{cd} \ \circ_{ab} \tag{6.56}$$

of maps $\mathcal{M}(S \sqcup \{a, b, c, d\}; g) \to \mathcal{M}(S; g + 2s)$.

(vi) For finite sets S_1, S_2, symbols a, b, c, d, and genera $g_1, g_2 \in \mathbb{A}$, one has the equality

$$\circ_{ab} \ {}_c\circ_d = \circ_{cd} \ {}_a\circ_b \tag{6.57}$$

of maps $\mathcal{M}(S_1 \sqcup \{a, c\}; g_1) \otimes \mathcal{M}(S_2 \sqcup \{b, d\}; g_2) \to \mathcal{M}(S_1 \sqcup S_2; g_1+g_2+s)$.

(vii) For finite sets S_1, S_2, symbols a, b, u, v, and genera $g_1, g_2 \in \mathbb{A}$, one has the equality

$$ {}_a\circ_b \ (\circ_{uv} \otimes \mathbb{1}) = \circ_{uv} \ {}_a\circ_b \tag{6.58}$$

of maps $\mathcal{M}(S_1 \sqcup \{a, u, v\}; g_1) \otimes \mathcal{M}(S_2 \sqcup \{b\}; g_2) \to \mathcal{M}(S_1 \sqcup S_2; g_1+g_2+s)$.

Notice that the ${}_u\circ_v$-operations preserve the $\mathbb{A}$-grading, while the contractions $\circ_{uv}$ raise it by the step s. The existing definitions of modular operads as given, e.g., in [3] have always assumed that $\mathbb{A} := \mathbb{N}$ and $s = 1$. There are several situations where this assumption is too restrictive.

Consider, for instance, two modular operads $\mathcal{M}'$ and $\mathcal{M}''$ with $\mathbb{A} := \mathbb{N}$ and $s = 1$. There exists an obvious product formula for the ${}_u\circ_v$- and $\circ_{uv}$-operations on the modular module $\mathcal{M} := \mathcal{M}' \otimes \mathcal{M}''$ with

$$\mathcal{M}(S; g) := \bigoplus_{g'+g''=g} \mathcal{M}'(S; g') \otimes \mathcal{M}(S; g'')$$

using those of $\mathcal{M}'$ resp. $\mathcal{M}''$, but the contractions thus defined clearly raise the genus grading by 2. To have a monoidal structure on the category of modular operads we must thus allow arbitrary steps. The products of operads $\mathcal{M}'$ and $\mathcal{M}''$ with the steps s' and s'', respectively, are then a modular operad with the step $s' + s''$. We will see later that assuming $\mathbb{A} = \mathbb{N}$ is also too restrictive.

Example 6.23 In all interesting examples of modular operad one has $s \neq 0$. It is an easy exercise that then a modular operad for which $\mathcal{M}(S; g) \neq 0$ only if card$(S) = 2$ and $g = 0$ is precisely a graded (non-unital) associative algebra A with an involution $\tau : A \to A$ such that $\tau(ab) = \tau(b)\tau(a)$ for all $a, b \in A$.

In the seminal paper [3] where modular operads were introduced, the following property of modular operads was always assumed.

Definition 6.17 A modular operad with $\mathbb{A} = \mathbb{N}$ and $s = 1$ is *stable* if

$$\mathcal{M}(S; g) = 0 \ \text{ for card}(S) \le 2,\ g = 0 \text{ and for card}(S) = 0,\ g = 1. \tag{6.59}$$

Using the notation

$$\mathfrak{S} := \big\{(S, g) \mid g \ge 2,\ \text{ or } g = 1 \text{ and card}(S) \ge 1,\ \text{ or } g = 0 \text{ and card}(S) \ge 3\big\}, \tag{6.60}$$

the stability of a modular operad $\mathcal{M}$ can be expressed by witting

$$\mathcal{M} = \{\mathcal{M}(S; g) \in \texttt{Chain} \mid (S, g) \in \mathfrak{S}\}.$$

Example 6.24 As cyclic operads, also modular operads exist in an arbitrary symmetric monoidal category, e.g., in the cartesian monoidal category $\texttt{Set}$ of sets. One has the terminal stable $\texttt{Set}$-modular operad $*_{\mathrm{Mod}}$ defined by

$$*_{\mathrm{Mod}}(S; g) := \begin{cases} * & \text{if } (S, g) \in \mathfrak{S}, \text{ and} \\ \emptyset & \text{otherwise,} \end{cases}$$

where $*$ is a chosen one-point set. The terminal stable modular operad $*_{\mathrm{Mod}}$ has a geometric interpretation extending the interpretation of the terminal cyclic operad given in Example 6.3.

Namely, for $(S, g) \in \mathfrak{S}$ consider the set $M(S; g)$ of isomorphism classes of oriented closed surfaces of genus g with holes indexed by S. It is a stable modular operad with ${}_a\circ_b$ as in Example 6.3, while $\circ_{uv}$ is given by attaching a handle as follows.

Let P be a surface of genus g with holes labeled by $S \sqcup \{u, v\}$. We let $\circ_{uv}(P)$ to denote the surface obtained from P by adding a tube connecting the circumference of the hole labeled u with the circumference of the hole labeled v. This gives rise to an operation

$$\circ_{ab} : M\big(S \sqcup \{u, v\}; g\big) \to M(S; g + 1)$$

on the set of isomorphism classes. Since there is for $(S, g) \in \mathfrak{S}$ only one isomorphism class in $M(S; g)$, one sees that M is isomorphic to the terminal stable modular operad $*_{\mathrm{Mod}}$.

Example 6.25 Property (6.59) is the abstraction of the stability of complex curves. A *stable curve* with marked points is a connected complex projective curve P whose only singularities are ordinary double points (nodal singularities), together with a "marking" given by an embedding of a set S into the set of smooth points of P. The stability means that there are no infinitesimal automorphisms of P fixing the marked and double points. Equivalently, each smooth component of P isomorphic to the complex projective space $\mathbb{CP}^1$ has at least three special points and each smooth component isomorphic to the torus has at least one special point, where a special point is either a double point or a marked point.

The *dual graph* $\Delta = \Delta(P)$ of a stable curve P is a labeled graph whose vertices are the components of P, edges are the nodes and its legs are the elements of S. An edge e_y corresponding to a nodal point y joins the vertices corresponding to the components intersecting at y. The vertex v_K corresponding to a branch K is labeled by the genus of the normalization of K. The construction of $\Delta(P)$ from a curve P is visualized in Fig. 6.5 taken from [12].

Let us denote by $\overline{\mathscr{M}}(S; g)$ the coarse moduli space [5, p. 347] of marked curves P whose dual graph $\Delta(P)$ has genus g. Obviously,

$$\overline{\mathscr{M}} = \left\{\overline{\mathscr{M}}(S; g) \mid (S, g) \in \mathtt{Cor} \times \mathbb{N}\right\}$$

is a modular module in the category of projective varieties. Since there are no stable curves of genus g if $2(g-1) + \mathrm{card}(S) \le 0$, $\overline{\mathscr{M}}$ automatically satisfies the stability condition (6.59).

For stable curves $P_1 \in \overline{\mathscr{M}}(S_1 \sqcup \{a\}; g_1)$ and $P_2 \in \overline{\mathscr{M}}(S_2 \sqcup \{b\}; g_2)$ we define

$$P_1 \,{}_a\!\circ_b\, P_2 \in \overline{\mathscr{M}}(S_1 \sqcup S_2; g_1 + g_2)$$

to be the curve obtained by the identification of the point $a \in P_1$ with $b \in P_2$ introducing a nodal singularity. The contraction $\circ_{uv}(P) \in \overline{\mathscr{M}}(S; g+1)$ of a curve $P \in \overline{\mathscr{M}}(S \sqcup \{u, v\}; g)$ is defined similarly. With these operations, $\overline{\mathscr{M}}$ is a modular operad in the category of complex projective varieties, with $\mathbb{A} = \mathbb{N}$ and $s = 1$.

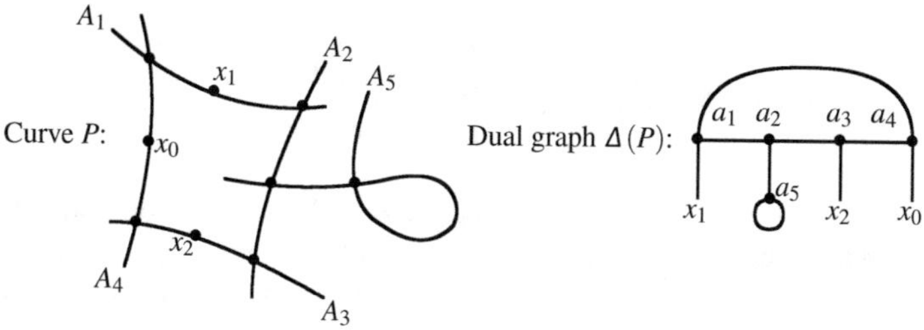

Fig. 6.5 A stable curve and its dual graph. The curve P on the left has five components, A_1, A_2, A_3, A_4, and A_5, and points marked by $S = \{x_0, x_1, x_2\}$. The dual graph $\Delta(P)$ on the right has five vertices a_1, a_2, a_3, a_4, and a_5 corresponding to the components of the curve and three legs labeled by the marked points

Informally, cyclic operads are modular operads without the contractions and the genus grading. One easily sees that the genus 0 part

$$\Box\mathscr{M} := \{\mathscr{M}(S;0) \mid S \in \mathtt{Cor}\}$$

of a modular operad $\mathscr{M}$ with the restricted ${}_a\circ_b$-operations forms a cyclic operad. One therefore has the forgetful functor

$$\Box : \mathtt{ModOp} \to \mathtt{CycOp}$$

from the category ModOp of modular operads with $\mathbb{A} = \mathbb{N}$ and the step $s = 1$ to the category of cyclic operads. It can be proved that it has a left adjoint

$$\mathrm{Mod} : \mathtt{CycOp} \to \mathtt{ModOp}$$

introduced in [8]. The functor Mod(−) preserves the stability.

Definition 6.18 The modular operad $\mathrm{Mod}(\mathscr{P})$ is the *modular completion* or the *modular envelope* of the cyclic operad $\mathscr{P}$.

Example 6.26 The terminal stable modular Set-operad $*_{\mathrm{Mod}}$ discussed in Example 6.24 is the modular completion of the terminal stable cyclic operad $*_{cyclic}$ from Example 6.2, i.e., one has the isomorphism

$$*_{\mathrm{Mod}} \cong \mathrm{Mod}(*_{cyclic})$$

of modular Set-operads (with $\mathbb{A} = \mathbb{N}$ and $s = 1$). Equivalently, the stable modular operad M of oriented surfaces of arbitrary genus is the modular completion of the stable cyclic operad M_0 of oriented surfaces of genus 0. The modular completion has in this case a clear geometric meaning.

Recall that $\mathscr{C}om \cong \mathrm{Span}(*_{cyclic})$ by (6.12). We conclude that the modular completion $\mathrm{Mod}(\mathscr{C}om)$ of the cyclic Chain-operad $\mathscr{C}om$ is the linearization of the terminal modular set-operad $*_{\mathrm{Mod}}$. Explicitly,

$$\mathrm{Mod}(\mathscr{C}om)(S;g) = \begin{cases} \Bbbk & \text{if } (S,g) \in \mathfrak{S}, \text{ and} \\ 0 & \text{otherwise,} \end{cases}$$

with all structure operations the canonical isomorphisms. The operad $\mathrm{Mod}(\mathscr{C}om)$ will play the fundamental rôle in our description of the algebraic structure of closed field theory given in Sect. 8.2 where it will be denoted $\mathscr{QC}$ and called the *quantum-closed operad*.

Example 6.27 In this example we discuss the stable modular operad $\mathscr{QO}$ consisting of homeomorphism classes of connected compact two-dimensional orientable surfaces with labeled marked points on the boundary. Stability means that we exclude surfaces of genus 0 with either one boundary component and less that three marked points, or two boundary components and no marked point. We also assume that all surfaces have at least one boundary component.

To define the operadic composition, it is convenient to replace each marked point by an interval embedded in the boundary and then glue one edge of a short strip to the interval. The edge of the strip opposite to the one glued to the interval is called an open end (of an open string). The surface(s) can be glued along the open ends where we allow only gluing resulting in orientable surfaces.

It is clear that the homeomorphism class of such a surface is determined by its genus, by the number of boundary components, and by the cyclically ordered sets of open ends at each boundary component. The modular operad $\mathcal{QO}$ therefore admits a purely combinatorial description given in the rest of this example. We will need the following notion.

Definition 6.19 A *cycle in a set* S is an equivalence class $(\!(x_1, \ldots, x_n)\!)$ of an n-tuple $(x_1, \ldots, x_n)$ of several distinct elements of S under the equivalence

$$(x_1, \ldots, x_n) \sim \tau(x_1, \ldots, x_n),$$

where $\tau \in \Sigma_n$ is the cyclic permutation given by $\tau(i) := i + 1$ for $1 \leq i \leq n - 1$, and $\tau(n) := 1$. In other words,

$$(\!(x_1, \ldots, x_n)\!) = \cdots = (\!(x_{n-i+1}, \ldots, x_n, x_1, \ldots, x_{n-i})\!) = \cdots = (\!(x_2, \ldots, x_n, x_1)\!).$$

We call n the *length* of the cycle.

We also admit the empty cycle $(\!(\,)\!)$, which is a cycle in any set. For a bijection $\rho : S \xrightarrow{\cong} T$ and a cycle $(\!(x_1, \ldots, x_n)\!)$ in S, define an induced cycle in T by

$$\rho\, (\!(x_1, \ldots, x_n)\!) := (\!(\rho(x_1), \ldots, \rho(x_n))\!).$$

We are going to introduce a stable modular operad

$$\mathcal{QO} = \left\{\mathcal{QO}(S; G) \in \mathtt{Set} \mid (S, G) \in \mathfrak{S}\right\}$$

in the category of sets which will be the combinatorial model of the stable part of the operad $\mathcal{QO}$.[10] Since the operadic genus of elements of $\mathcal{QO}$ does not coincide with the geometric genus of the surface it represents, we denoted it in this particular example by the capital G instead of g that we used for the operadic genus before.

[10]The set $\mathfrak{S}$ of stable pairs was defined in (6.60).

The component $\mathcal{QO}(S; G)$ is defined as

$$\mathcal{QO}(S; G) := \Big\{\{\mathbf{c}_1, \dots, \mathbf{c}_b\}^g \mid b > 0,\ g \geq 0,\ \bigsqcup_{i=1}^{b} \mathbf{c}_i = S,\ G = 2g + b - 1\Big\},$$

where $\{\mathbf{c}_1, \dots, \mathbf{c}_b\}^G$ is a symbol consisting of a non-negative integer $g \geq 0$ and an (unordered) set of cycles whose disjoint union is S. The natural number g is the geometric genus of the surface represented by a given symbol. For a bijection $\rho : S \xrightarrow{\cong} T$, let

$$\mathcal{QO}(\rho)(\{\mathbf{c}_1, \dots, \mathbf{c}_b\}^g) := \{\rho(\mathbf{c}_1), \dots, \rho(\mathbf{c}_b)\}^g.$$

Next, we define the composition operations

$${}_a\circ_b : \mathcal{QO}\big(S_1 \sqcup \{a\}, G_1\big) \otimes \mathcal{QO}\big(S_2 \sqcup \{b\}, G_2\big) \to \mathcal{QO}\big(S_1 \sqcup S_2, G_1 + G_2\big).$$

Assume that $\mathbf{c}_i = (\!(a, x_1, \dots, x_m)\!)$ is a cycle in $S_1 \sqcup \{a\}$ and let $\mathbf{d}_j = (\!(b, y_1, \dots, y_n)\!)$ be a cycle in $S_2 \sqcup \{b\}$. Then

$$\begin{aligned}&\{\mathbf{c}_1, \dots, \mathbf{c}_{b_1}\}^{g_1}\ {}_a\circ_b \{\mathbf{d}_1, \dots, \mathbf{d}_{b_2}\}^{g_2} \\ &:= \big\{ (\!(x_1, \dots, x_m, y_1, \dots, y_n)\!), \mathbf{c}_1, \dots, \widehat{\mathbf{c}_i}, \dots, \mathbf{c}_{b_1}, \mathbf{d}_1, \dots, \widehat{\mathbf{d}_j}, \dots, \mathbf{d}_{b_2}\big\}^{g_1+g_2}.\end{aligned}$$

The contractions

$$\circ_{uv} = \circ_{vu} : \mathcal{QO}\big(S \sqcup \{u, v\}; G\big) \to \mathcal{QO}(S; G + 1)$$

are defined as follows. Let $\{\mathbf{c}_1, \dots, \mathbf{c}_b\}^g \in \mathcal{QO}(S, G)$. If there are $i < j$ such that $\mathbf{c}_i = (\!(u, x_1, \dots, x_m)\!)$ and $\mathbf{c}_j = (\!(v, y_1, \dots, y_n)\!)$, then define

$$\circ_{uv}(\{\mathbf{c}_1, \dots, \mathbf{c}_b\}^g) := \{(\!(x_1, \dots, x_m, y_1, \dots, y_m)\!), \mathbf{c}_1, \dots, \widehat{\mathbf{c}_i}, \dots, \widehat{\mathbf{c}_j}, \dots, \mathbf{c}_b\}^{g+1}.$$

Otherwise, there is i such that $\mathbf{c}_i = (\!(u, x_1, \dots, x_m, v, y_1, \dots, y_n)\!)$. Then define

$$\circ_{uv}(\{\mathbf{c}_1, \dots, \mathbf{c}_b\}^g) := \{(\!(x_1, \dots, x_m)\!), (\!(y_1, \dots, y_n)\!), \mathbf{c}_1, \dots, \widehat{\mathbf{c}_i}, \dots, \mathbf{c}_b\}^g.$$

Notice that we allow repeated empty cycles to appear in $\{\mathbf{c}_1, \dots, \mathbf{c}_b\}^g$, for example, $\{(\!()\!), (\!()\!), (\!(3)\!), (\!(14)\!), (\!(25)\!)\}^2 \in \mathcal{QO}([5], 8)$. Also notice that $\circ_{uv}$ can produce empty cycles, e.g.,

$$\circ_{uv}\{(\!(u)\!), (\!(v)\!)\}^g = \{(\!()\!)\}^{g+1} \text{ and } \circ_{uv}\{(\!(uv)\!)\}^g = \{(\!()\!), (\!()\!)\}^g.$$

Observe that

$$\mathrm{card}(\mathscr{QO}(S, G)) < \infty$$

for any $(S, G) \in \mathfrak{G}$. The reader can easily verify that:

Theorem 6.1 *$\mathscr{QO}$ is a stable modular operad in the category of sets.*

As shown in [10], $\mathscr{QO}$ is the modular completion of the non-Σ cyclic operad $\underline{\mathscr{A}ss}$ from Example 6.17 in a certain category of non-Σ modular operads. The relation between $\mathscr{QO}$ and $\underline{\mathscr{A}ss}$ is therefore similar as the relation between $\mathscr{QC}$ and $\mathscr{C}om$ described in Example 6.26.

Example 6.28 Let V be a graded vector space and $s \in V \otimes V$ a symmetric degree 0 tensor. We are going to define a modular extension of the cyclic endomorphism operad from Example 6.6. For a finite set S and $g \in \mathbb{A}$ we put

$$\mathscr{E}nd_V(S; g) := Lin\big(\textstyle\bigotimes_{c\in S} V_c, \Bbbk\big) = \big(\textstyle\bigotimes_{c\in S} V_c\big)^{\#} \tag{6.61}$$

so $\mathscr{E}nd_V(S; g)$ equals the component $\mathscr{E}nd_V(S)$ of the cyclic endomorphism operad for each $g \in \mathbb{A}$. The action of isomorphisms of finite sets and also the ${}_a\circ_b$-operations are defined precisely as in Example 6.6. The contraction $\circ_{uv} f \in \mathscr{E}nd_V(S; g+s)$ of $f \in \mathscr{E}nd_V\big(S \sqcup \{u, v\}; g\big)$ is the composition

$$\bigotimes_{c\in S} V_c \cong \Bbbk \otimes \bigotimes_{c\in S} V_c \xrightarrow{s\otimes \mathbb{1}} V_u \otimes V_v \otimes \bigotimes_{c\in S} V_c \xrightarrow{\cong} \bigotimes_{c\in S\sqcup\{u,v\}} V_c \xrightarrow{f} \Bbbk, \tag{6.62}$$

where s is interpreted as a degree 0 map $\Bbbk \to V_u \otimes V_v$. It follows from the commutativity of the diagram

$$\begin{array}{ccccc}
 & \Bbbk \otimes \bigotimes_{c\in S} V_c & \xrightarrow{s\otimes \mathbb{1}} & V_u \otimes V_v \otimes \bigotimes_{c\in S} V_c & \\
\cong\nearrow & & & & \searrow\cong \\
\bigotimes_{c\in S} V_c & & & & \bigotimes_{c\in S\sqcup\{u,v\}} V_c \\
\cong\searrow & & & & \nearrow\cong \\
 & \bigotimes_{c\in S} V_c \otimes \Bbbk & \xrightarrow[\mathbb{1}\otimes s]{} & \bigotimes_{c\in S} V_c \otimes V_u \otimes V_v &
\end{array} \tag{6.63}$$

that we could replace $s \otimes \mathbb{1}$ in (6.62) by $\mathbb{1} \otimes s$ with the same result. Equivalently, one may define $\circ_{uv} f$ as the result of the application

$$\Big(\bigotimes_{c\in S\sqcup\{u,v\}} V_c\Big)^{\#} \xrightarrow{\cong} \Big(V_u \otimes V_v \otimes \bigotimes_{c\in S} V_c\Big)^{\#} \xrightarrow{(s\otimes\mathbb{1})^{\#}} \Big(\Bbbk \otimes \bigotimes_{c\in S} V_c\Big)^{\#} \cong \Big(\bigotimes_{c\in S} V_c\Big)^{\#} \tag{6.64}$$

to $f \in \big(\bigotimes_{c\in S\sqcup\{u,v\}} V_c\big)^{\#}$. In shorthand, $\circ_{uv} f := \big(s \otimes \mathbb{1}_S\big)^{\#}(f)$. It is not difficult to prove that the modular module

$$\mathcal{E}nd_V = \big\{\mathcal{E}nd_V(S; g) \,|\, (S; g) \in \mathtt{Cor} \times \mathbb{A}\big\}$$

is a modular operad.[11] The modular version of the operad $\mathcal{D}ne_V$ of Example 6.7 can be constructed similarly. Neither $\mathcal{E}nd_V$ or $\mathcal{D}ne_V$ are stable.

As expected, modular endomorphisms operads are used to define algebras over modular operads:

Definition 6.20 Let V be a graded vector space and $s \in V \otimes V$ a symmetric degree 0 tensor. An *algebra* over a modular operad $\mathcal{M}$, or an *$\mathcal{M}$-algebra*, is a morphism $\alpha : \mathcal{M} \to \mathcal{E}nd_V$ of modular operads.

It follows from the universal property defining the modular completion of a cyclic operad that the category of algebras over a cyclic operad $\mathcal{P}$ is isomorphic to the category of algebras over its modular completion $\mathrm{Mod}(\mathcal{P})$.

Forgetting the structure operations ${}_a\circ_b$ and $\circ_{uv}$ of modular operads induces the functor

$$\square : \mathtt{ModOp} \to \mathtt{ModMod}$$

from the category of modular operads to the category of modular modules. It has a left adjoint

$$\mathbb{F} : \mathtt{ModMod} \to \mathtt{ModOp}.$$

The operad $\mathbb{F}(E)$ is the *free modular operad* generated by the modular module E. It is characterized by the existence of a natural isomorphism of morphism spaces

$$\mathtt{ModMod}\big(E, \square(\mathcal{M})\big) \cong \mathtt{ModOp}\big(\mathbb{F}(E), \mathcal{M}\big)$$

[11] So we denote both the cyclic endomorphism operad and its modular version by the same symbol. The meaning will however always be clear from the context.

which can easily be converted into a diagram analogous to the one in Proposition 6.2.

Free modular operads admit a description parallel to free cyclic operads given in Sect. 6.1. Let us firstly introduce some necessary notions.

Definition 6.21 A *labeled graph* is a couple $\boldsymbol{\Gamma} = (\Gamma, \ell)$ consisting of a graph Γ as in Definition 6.8 and a *labeling* $\ell : \mathit{Vert}(\Gamma) \to \mathbb{A}$. The *genus* $g(\boldsymbol{\Gamma}) \in \mathbb{A}$ of a labeled graph is defined by the formula

$$g(\boldsymbol{\Gamma}) := s \cdot b_1(\Gamma) + \sum_{v \in \mathit{Vert}(\Gamma)} \ell(v),$$

where $b_1(\Gamma)$ is the first Betti number of the geometric realization of Γ, i.e., the number of independent circuits of Γ.

An *isomorphism* of $\boldsymbol{\phi} : (\Gamma_0, \ell_0) \to (\Gamma_1, \ell_1)$ of labeled graphs is an isomorphism $\phi : \Gamma_0 \xrightarrow{\cong} \Gamma_1$ of the underlying graphs compatible with the labelings. The grafting extends to labeled graphs in the straightforward manner, namely

$$\boldsymbol{\Gamma}_1 \,{}_a\circ_b\, \boldsymbol{\Gamma}_2 = (\Gamma_1 \,{}_a\circ_b\, \Gamma_2; \ell_1 \,{}_a\circ_b\, \ell_2), \tag{6.65}$$

where $\Gamma_1 \,{}_a\circ_b\, \Gamma_2$ is as in Definition 6.9 and

$$(\ell_1 \,{}_a\circ_b\, \ell_2)|_{\mathit{Vert}(\Gamma_i)} := \ell_i \ \text{ for } i = 1, 2.$$

One clearly has

$$g(\boldsymbol{\Gamma}_1 \,{}_a\circ_b\, \boldsymbol{\Gamma}_2) = g(\boldsymbol{\Gamma}_1) + g(\boldsymbol{\Gamma}_2).$$

Suppose that $\boldsymbol{\Gamma} = (\Gamma, \ell)$ is a labeled graph with $\mathit{Leg}(\Gamma) = S \sqcup \{u, v\}$. We define the contracted labeled graph $\circ_{uv}\boldsymbol{\Gamma} = (\circ_{uv}\Gamma, \circ_{uv}\ell)$ as follows. The graph $\circ_{uv}\Gamma$ has the same set of flags and its partition as Γ, in particular, $\mathit{Vert}(\circ_{uv}\Gamma) = \mathit{Vert}(\Gamma)$. The involution $\circ_{uv}(\sigma)$ of $\mathit{Flag}(\circ_{uv}\Gamma)$ agrees with the involution of Γ on $\mathit{Flag}(\Gamma) \backslash \{u, v\}$ while $\circ_{uv}(\sigma)(u) = v$. Informally, $\circ_{uv}\Gamma$ is obtained from Γ by connecting the free ends of the legs u and v, creating a loop, as expressed in the schematic picture

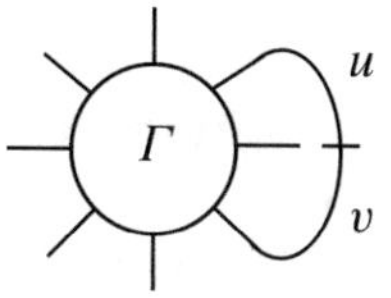

The labeling $\circ_{uv}\ell$ is given as the composition $Vert(\circ_{uv}\Gamma) = Vert(\Gamma) \xrightarrow{\ell} \mathbb{N}$. Notice that

$$g(\circ_{uv}\boldsymbol{\Gamma}) = g(\boldsymbol{\Gamma}) + s.$$

Definition 6.22 We call the labeled graph $\circ_{uv}\boldsymbol{\Gamma}$ the *contraction* of $\boldsymbol{\Gamma}$.

For a labeled graph $\boldsymbol{\Gamma}$ and a modular module E we consider an analog

$$E(\boldsymbol{\Gamma}) := \bigotimes_{v \in Vert(\Gamma)} E\big(Leg(v); \ell(v)\big) \tag{6.66}$$

of the vector space (6.33). Each graph isomorphism $\phi : \boldsymbol{\Gamma}_0 \xrightarrow{\cong} \boldsymbol{\Gamma}_1$ of labeled graphs clearly induces an isomorphism

$$E(\phi) : E(\boldsymbol{\Gamma}_0) \xrightarrow{\cong} E(\boldsymbol{\Gamma}_1)$$

of the spaces (6.66). Mimicking (6.34), we define for a finite set S and a genus $g \in \mathbb{A}$

$$\mathbb{F}(E)(S; g) := \frac{\bigoplus_G E(\boldsymbol{\Gamma})}{\sim} \tag{6.67}$$

with the sum taken over all labeled graphs $\boldsymbol{\Gamma} = (\Gamma, \ell)$ having $Leg(\Gamma) = S$ and $g(\boldsymbol{\Gamma}) = g$. The relation $\sim$ identifies $x \in E(\boldsymbol{\Gamma}_0)$ with its image $E(\phi)(x) \in E(\boldsymbol{\Gamma}_1)$ for any isomorphism $\phi : \boldsymbol{\Gamma}_0 \to \boldsymbol{\Gamma}_1$ of labeled graphs that induces the identity map of the set of the legs.

Proposition 6.9 *The modular module* $\mathbb{F}(E) = \big\{\mathbb{F}(E)(S; g) \mid S \in \mathtt{Cor},\ g \in \mathbb{A}\big\}$ *is a modular operad.*

Proof. The actions (6.48) and compositions (6.51) are defined as in the proof of Proposition 6.1; we leave the details for the reader. To define the $\circ_{uv}$-operations, we recall the contraction of Definition 6.22 and notice the canonical isomorphism $E(\boldsymbol{\Gamma}) \cong E(\circ_{uv}\boldsymbol{\Gamma})$ which induces isomorphisms of the quotients

$$\circ_{uv} : \mathbb{F}(E)\big(S \sqcup \{u, v\}; g\big) \cong \mathbb{F}(E)\big(S; g + s\big).$$

The axioms of modular operads are easy to verify.

It is simple to see that, if $\mathbb{A} = \mathbb{N}$ and $s = 1$, the modular operad $\mathbb{F}(E)$ is stable if and only if $E(S; g) \neq 0$ implies $(S, g) \in \mathfrak{S}$. The following analog of Proposition 6.2 holds.

Proposition 6.10 *The modular operad $\mathbb{F}(E)$ is the free modular operad on the modular module E.*

We finish this section by translating Definition 6.16 of modular operads into the skeletal language. Recall that $[n] := \{1, \ldots, n\}$, with $[0]$ the empty set $\emptyset$. For $\mathcal{M}$ as in (6.50), $n \geq 0$ and $g \in \mathbb{A}$ denote $\mathcal{M}(n; g) := \mathcal{M}([n]; g)$. The operations

$$_i\bar{\circ}_j : \mathcal{M}(m+1; g_1) \otimes \mathcal{M}(n+1; g_2) \to \mathcal{M}(m+n; g_1+g_2)$$

are, for $1 \leq i \leq m+1$, $1 \leq j \leq n+1$ and $g_1, g_2 \in \mathbb{A}$, given by the obvious analog of formula (6.8). To define the skeletal version

$$\bar{\circ}_{ij} : \mathcal{M}(n+2; g+s) \to \mathcal{M}(n; g), \ 1 \leq i, j \leq n+2, \ g \in \mathbb{A}, \tag{6.68}$$

of the contractions $\circ_{uv}$ in (6.52), we need an auxiliary map

$$\tau = \tau_{ij} : [n+2] \setminus \{i, j\} \to [n] \tag{6.69}$$

given by

$$\tau_{ij}(a) := \begin{cases} a, & \text{for } 1 \leq a < i, \\ a-1, & \text{for } i < a < j, \text{ and} \\ a-2, & \text{for } j < a \leq n+2, \end{cases}$$

if $i < j$, while for $i > j$ we set $\tau_{ij} := \tau_{ji}$. Then $\bar{\circ}_{ij}$ in (6.68) is the composition

$$\mathcal{M}(n+2; g+s) \xrightarrow{\circ_{ij}} \mathcal{M}\big([n+2] \setminus \{i, j\}; g\big) \xrightarrow{\mathcal{M}(\tau ij)} \mathcal{M}(n; g), \tag{6.70}$$

where $\circ_{ij}$ is the contraction (6.52) with $S = [n+2]$. An involved but straightforward calculation shows that the above structure has the properties listed in the following

Definition 6.23 A *modular operad* $\mathcal{M}$ is a family

$$\mathcal{M} = \{\mathcal{M}(n; g) \mid n \geq 0, \ g \in \mathbb{A}\}$$

of dg-vector spaces together with linear left actions

$$\Sigma_n \times \mathcal{M}(n; g) \to \mathcal{M}(n; g), \ n \geq 1, \ g \in \mathbb{A},$$

of the symmetric groups Σ_n, degree 0 morphisms

$$_i\bar{\circ}_j : \mathcal{M}(m+1; g_1) \otimes \mathcal{M}(n+1; g_2) \to \mathcal{M}(m+n; g_1+g_2),$$

defined for $m, n \geq 0$, $1 \leq i \leq m+1$, $1 \leq j \leq n+1$, $g_1, g_2 \in \mathbb{A}$, and degree 0 morphisms (the contractions)

$$\overline{\circ}_{ij} = \overline{\circ}_{ji} : \mathcal{M}(n+2; g+s) \to \mathcal{M}(n; g)$$

defined for $n \geq 0$, $1 \leq i \neq j \leq n$ and $g \in \mathbb{A}$. These data are required to satisfy the obvious modular versions of axioms (i)–(iii) of Definition 6.4 involving the ${}_i\overline{\circ}_j$-operations, plus the following ones.

(i) For each $n \geq 0$, $g \in \mathbb{A}$, $x \in \mathcal{M}(n+2; g)$ and a permutation $\rho \in \Sigma_{n+2}$,

$$\overline{\circ}_{\rho(i)\rho(j)}(\rho x) = \lambda \overline{\circ}_{ij}(x),$$

where $\lambda \in \Sigma_n$ is the composition

$$[n] \xrightarrow{\tau_{ij}^{-1}} [n+2] \setminus \{i, j\} \xrightarrow{\rho} [n+2] \setminus \{\rho(i), \rho(j)\} \xrightarrow{\tau_{\rho(i)\rho(j)}} [n].$$

(ii) For $m \geq 0$, $g \in \mathbb{A}$, $x \in \mathcal{M}(m+4; g)$, $1 \leq c < d \leq m+4$ and $1 \leq a < b \leq m+2$,

$$\overline{\circ}_{ab}\overline{\circ}_{cd}(x) = \begin{cases} \overline{\circ}_{c-2,d-2}\overline{\circ}_{ab}(x), & \text{if } b < c, \\ \overline{\circ}_{c-1,d-2}\overline{\circ}_{a,b+1}(x), & \text{if } a < c \leq b < d-1, \\ \overline{\circ}_{c-1,d-1}\overline{\circ}_{a,b+2}(x), & \text{if } a < c,\ d-1 \leq b, \\ \overline{\circ}_{c,d-2}\overline{\circ}_{a+1,b+1}(x), & \text{if } c \leq a < b < d-1, \\ \overline{\circ}_{c,d-1}\overline{\circ}_{a+1,b+2}(x), & \text{if } c \leq a < d-1 \leq b, \text{ and} \\ \overline{\circ}_{c,d}\overline{\circ}_{a+2,b+2}(x), & \text{if } d-1 \leq a. \end{cases}$$

(iii) For $m, n \geq 0$, $g_1, g_2 \in \mathbb{A}$, $x \in \mathcal{M}(m+2; g_1)$, $y \in \mathcal{M}(n+2; g_2)$, $1 \leq a \leq c-1$ and $c \leq b \leq c+n$,

$$\overline{\circ}_{ab}(x\ {}_c\overline{\circ}_d\ y)$$
$$= \begin{cases} \overline{\circ}_{c+n,a-b+c+n}(x\ {}_a\overline{\circ}_{b+d-c+1}\ y), & \text{if } c \leq b < c-d+n+2, \text{ and} \\ \overline{\circ}_{c+n,a-b+c+n}(x\ {}_a\overline{\circ}_{b+d-c-n-1}\ y), & \text{if } c-d+n+2 \leq b \leq c+n. \end{cases}$$

If $c+n+1 \leq a \leq m+n+2$ and x, y, b are as above, then

$$\overline{\circ}_{ab}(x\ {}_c\overline{\circ}_d\ y)$$
$$= \begin{cases} \overline{\circ}_{c,a-b+c}(x\ {}_{a-n}\overline{\circ}_{b+d-c+1}\ y), & \text{if } c \leq b < c-d+n+2, \text{ and} \\ \overline{\circ}_{c,a-b+c}(x\ {}_{a-n}\overline{\circ}_{b+d-c-n-1}\ y), & \text{if } c-d+n+2 \leq b \leq c+n. \end{cases}$$

(iv) For $m, n \geq 0$, $g_1, g_2 \in \mathbb{A}$, $x \in \mathscr{M}(m+3; g_1)$, $x \in \mathscr{M}(n+1; g_2)$,

$$\overline{\circ}_{ab}(x\ {}_c\overline{\circ}_d\ y)$$
$$= \begin{cases} \overline{\circ}_{ab}(x)\ {}_{c-2}\overline{\circ}_d\ y, & \text{if } 1 \leq a < b < c, \\ \overline{\circ}_{a,b-n+1}(x)\ {}_{c-1}\overline{\circ}_d\ y, & \text{if } 1 \leq a < c,\ c+n \leq b \leq m+n+2, \\ \overline{\circ}_{a-n+1,b-n+1}(x)\ {}_c\overline{\circ}_d\ y, & \text{if } c+n \leq a < b \leq m+n+2. \end{cases}$$

Denoting

$$\mathfrak{S}_{\rm sk} := \big\{(n, g) \mid g = 0 \text{ and } n \geq 3,\ \text{ or } g = 1 \text{ and } n \geq 1,\ \text{ or } g \geq 2\big\}, \tag{6.71}$$

one sees that a modular operad $\mathscr{M}$ is stable if and only if $\mathscr{M}(n; g) \neq 0$ implies $(n, g) \in \mathfrak{S}_{\rm sk}$.

Example 6.29 Extending the calculations of Example 6.8, one can easily describe the skeletal version of the modular endomorphism operad $\mathscr{E}nd_V$. One has

$$\mathscr{E}nd_V(n; g) := \mathscr{E}nd_V\big([n]; g\big) \cong Lin(V^{\otimes n}, \Bbbk),\ n \geq 0, g \in \mathbb{A}, \tag{6.72}$$

with the skeletal operations ${}_i\overline{\circ}_j$ given by (6.25). For

$$f \in \mathscr{E}nd_V(n+2; g+1) \cong Lin(V^{\otimes n+2}, \Bbbk),$$

$1 \leq i < j \leq n+2$ and homogeneous $v_1, \ldots, v_n \in V$, one obtains

$$\overline{\circ}_{ij} f(v_1, \ldots, v_n) = \sum (-1)^\kappa f(v_1, \ldots, v_{i-1}, s_i', v_i, \ldots, v_{j-2}, s_i'', v_{j-1}, \ldots, v_n) \tag{6.73}$$

with

$$\kappa = |s_i''|(|v_i| + \cdots + |v_{j-2}|). \tag{6.74}$$

Let us explain the sign. Denote

$$\omega_1 := v_1 \otimes \cdots \otimes v_{i-1} \in V_1 := V^{\otimes(i-1)},$$
$$\omega_2 := v_i \otimes \cdots \otimes v_{j-2} \in V_2 := V^{\otimes(j-i-1)}$$
$$\text{and } \omega_3 := v_{j-1} \otimes \cdots \otimes v_n \in V_3 := V^{\otimes(n-j+2)}.$$

The skeletal $\overline{\circ}_{ij} f$ is the composition of the canonical isomorphism

$$V_1 \otimes V_2 \otimes V_3 \xrightarrow{\cong} \Bbbk \otimes V_1 \otimes V_2 \otimes V_3 \tag{6.75}$$

followed by

$$(s \otimes \mathbb{1}) : \Bbbk \otimes V_1 \otimes V_2 \otimes V_3 \to V_u \otimes V_v \otimes V_1 \otimes V_2 \otimes V_3$$

composed with the permutation

$$\rho : V_u \otimes V_v \otimes V_1 \otimes V_2 \otimes V_3 \xrightarrow{\cong} V_1 \otimes V_u \otimes V_2 \otimes V_v \otimes V_3$$

and finally followed by

$$f : V_1 \otimes V_u \otimes V_2 \otimes V_v \otimes V_3 \longrightarrow \Bbbk.$$

Let us apply this composition on the element

$$v_1 \otimes \cdots \otimes v_n = \omega_1 \otimes \omega_2 \otimes \omega_3 \in V^{\otimes(n)}.$$

Isomorphism (6.75) brings $\omega_1 \otimes \omega_2 \otimes \omega_3$ into $1 \otimes \omega_1 \otimes \omega_2 \otimes \omega_3$. One then has

$$(s \otimes \mathbb{1})(1 \otimes \omega_1 \otimes \omega_2 \otimes \omega_3) = \sum s_i' \otimes s_i'' \otimes \omega_1 \otimes \omega_2 \otimes \omega_3,$$

while

$$\rho(s_i' \otimes s_i'' \otimes \omega_1 \otimes \omega_2 \otimes \omega_3) = (-1)^{|s_i'||\omega_2|} \omega_1 \otimes s_i' \otimes \omega_2 \otimes s_i'' \otimes \omega_3, \tag{6.76}$$

so the result is $\sum (-1)^{|s_i''||\omega_2|} f(\omega_1 \otimes s_i' \otimes \omega_2 \otimes s_i'' \otimes \omega_3)$ as claimed.

6.5 Odd Modular Operads

One of our fundamental constructions used in this work is the Feynman transform of a modular operads recalled below in Sect. 7.2. Quite surprisingly, the Feynman transform is not an ordinary modular operad, but its odd version.

Definition 6.24 An *odd modular operad*[12] with step s is a modular module

$$\mathscr{T} = \big\{\mathscr{T}(S; g) \in \mathtt{Chain};\ (S; g) \in \mathtt{Cor} \times \mathbb{A}\big\} \tag{6.77}$$

together with degree $+1$ morphisms (compositions)

$${}_a\bullet_b : \mathscr{T}\big(S_1 \sqcup \{a\}; g_1\big) \otimes \mathscr{T}\big(S_2 \sqcup \{b\}; g_2\big) \to \mathscr{T}(S_1 \sqcup S_2; g_1 + g_2) \tag{6.78}$$

[12]This terminology was introduced by Ralph Kaufmann; the name “twisted modular operad” is sometimes used, too.

defined for arbitrary finite disjoint sets S_1, S_2, symbols a, b, genera $g_1, g_2 \in \mathbb{A}$, and degree $+1$ contractions

$$\bullet_{uv} = \bullet_{vu} : \mathscr{T}\big(S \sqcup \{u, v\}; g\big) \to \mathscr{T}(S; g + s) \tag{6.79}$$

given for any finite set S, genus $g \in \mathbb{A}$, and symbols u, v. These data are required to satisfy axioms of Definition 6.16 for the operations ${}_a\circ_b$ and $\circ_{uv}$, with the only difference that the formulas in axioms (iv)–(vii) acquire the minus signs, i.e., read as

$$_a\bullet_b(\mathbb{1} \otimes {}_c\bullet_d) = -\, {}_c\bullet_d({}_a\bullet_b \otimes \mathbb{1}), \tag{6.80}$$

$$\bullet_{ab}\, \bullet_{cd} = -\, \bullet_{cd}\, \bullet_{ab}, \tag{6.81}$$

$$\bullet_{ab}\, {}_c\bullet_d = -\, \bullet_{cd}\, {}_a\bullet_b, \text{ and} \tag{6.82}$$

$$_a\bullet_b\, (\bullet_{uv} \otimes \mathbb{1}) = -\, \bullet_{uv}\, {}_a\bullet_b\,. \tag{6.83}$$

A *morphism* of odd modular operads is a morphism of the underlying modular modules commuting with all structure operations.

The minus signs in (6.80)–(6.83) are forced by the Koszul sign conventions, as both the compositions and contractions are "objects" of degree $+1$. Informally, odd modular operads are modular operads whose structure operations have "wrong" degrees and also the signs of some of the axioms are "wrong." Because of nontrivial signs and degrees, odd modular operads, unlike the ordinary ones, do not exist in an arbitrary symmetric monoidal category but, e.g., in symmetric monoidal categories enriched over graded vector spaces.

Remark 6.11 The category of *Lin* of $\mathbb{Z}$-graded vector spaces and their homogeneous linear maps of arbitrary degrees has two symmetric monoidal structures, the standard one and the one which we call, from reasons which will became clear later, the *Montreal* monoidal structure. The monoidal product of objects is for both structures the standard tensor product of graded vector spaces, but the structures differ by their actions on morphisms. The prevailing convention is that, for homogeneous maps $f : V' \to W'$, $g : V'' \to W''$ and homogeneous vectors $u \in V'$, $v \in W'$ one defines

$$(f \otimes g)(u \otimes v) = (-1)^{|g||u|} f(u) \otimes g(v), \tag{6.84}$$

while some categorists at McGill University in Montreal would prefer

$$(f \otimes g)(u \otimes v) = (-1)^{|f||v|} f(u) \otimes g(v). \tag{6.85}$$

The second convention would follow from the Koszul sign rule if we apply the morphisms from the right. Equation (6.85) then reads as

$$(u \otimes v)(f \otimes g) = (-1)^{|f||v|} f(u) \otimes g(v)$$

and the unexpected sign comes from the commuting f over v. We denote, only for the purposes of this remark, the first monoidal structure by $\otimes_S$ and second by $\otimes_M$ (K abbreviating standard and M Montreal). We denote the corresponding monoidal categories by Lin_S and Lin_M, respectively.

The two monoidal structures on Lin are related as follows. For any monoidal category $\mathtt{M}$ with the monoidal product $\odot$ one can equip the same $\mathtt{M}$ by the *opposite* monoidal structure $\odot^\dagger$ defined on objects by $A \odot^\dagger B := A \odot B$ and similarly on morphisms; let us denote $\mathtt{M}$ with the opposite monoidal structure by $\mathtt{M}^\dagger$. It turns out that the categories $Lin_M^\dagger$ and Lin_S are isomorphic. The isomorphism is the identity of the underlying categories, and the transformation turning $\otimes_M^\dagger$ into $\otimes_S$ is the family of maps

$$\Phi_{U,V} : U \otimes_M^\dagger V = V \otimes U \to U \otimes V = U \otimes_S V$$

given by $\Phi_{U,V}(v \otimes u) := v \otimes u$ for $u \in U$ and $v \in V$.[13]

The category Lin with both monoidal structures is enriched over itself. For homogeneous maps $f : A \to V$ and $g : W \to B$, the corresponding enriched functors

$$f^\# : Lin(V, W) \to Lin(A, W) \text{ and } g_\# : Lin(V, W) \to Lin(V, B)$$

are given, for $\varphi \in Lin(V, W)$, by

$$f^\#(\varphi) := (-1)^{|f||\varphi|} \varphi \circ f \text{ and } g_\#(\varphi) := \varphi \circ g.$$

One can easily check that the obvious canonical isomorphism

$$Lin\big(A \otimes B, C\big) \cong Lin\big(A, Lin(B, C)\big)$$

is functorial in B for both $\otimes = \otimes_S$ and $\otimes = \otimes_M$.

Since odd modular operads possess operations of odd degrees, the form of their axioms evaluated at concrete elements may depend on the chosen monoidal structure of Lin. Let us, for instance, evaluate axiom (6.80) at homogeneous elements

$$x \in \mathscr{T}\big(S_1 \sqcup \{a\}; g_1\big),\ y \in \mathscr{T}\big(S_2 \sqcup \{b, c\}; g_2\big) \text{ and } z \in \mathscr{T}\big(S_3 \sqcup \{d\}; g_3\big)$$

[13] Notice there are no signs!

i.e., calculate

$$ {}_a\bullet_b(\mathbb{1} \otimes {}_c\bullet_d)(x \otimes y \otimes z) = -\, {}_c\bullet_d({}_a\bullet_b \otimes \mathbb{1})(x \otimes y \otimes z). $$

While in Lin_S we get

$$ (-1)^{|x|} x \,{}_a\bullet_b (y \,{}_c\bullet_d\, z) = -(x \,{}_a\bullet_b\, y) \,{}_c\bullet_d\, z, \tag{6.86} $$

in Lin_M we obtain

$$ x \,{}_a\bullet_b (y \,{}_c\bullet_d\, z) = -(-1)^{|z|}(x \,{}_a\bullet_b\, y) \,{}_c\bullet_d\, z. \tag{6.87} $$

Likewise, axiom (6.83) in Lin_S reads

$$ \bullet_{uv}\,(x)\,{}_a\bullet_b\, y = -\;\bullet_{uv}\,(x\,{}_a\bullet_b\, y), \tag{6.88} $$

while in Lin_M one would get

$$ (-1)^{|y|}\;\bullet_{uv}\,(x)\,{}_a\bullet_b\, y = -\;\bullet_{uv}\,(x\,{}_a\bullet_b\, y) \tag{6.89} $$

for x, y belonging to the appropriate components of $\mathscr{T}$. Also the derivation property of the differential with respect to the ${}_a\bullet_b$-operation depends on the chosen monoidal structures. In Lin_S it reads

$$ d(x\,{}_a\bullet_b\, y) = -(dx)\,{}_a\bullet_b\, y - (-1)^{|x|} x\,{}_a\bullet_b (dy), $$

while in Lin_M it is

$$ d(x\,{}_a\bullet_b\, y) = -(-1)^{|y|}(dx)\,{}_a\bullet_b\, y - x\,{}_a\bullet_b (dy). $$

Axioms (6.81) and (6.82) are the same in both monoidal structures.

It fortunately turns out that the categories of odd modular operads in Lin_S and in Lin_S are isomorphic. Indeed, we leave as an exercise to prove that the modification

$$ x\,{}_a\bullet_b\, y \mapsto (-1)^{|x|+|y|} x\,{}_a\bullet_b\, y,\;\; \bullet_{uv}(x) \mapsto (-1)^{|x|}\;\bullet_{uv}\,(x),\;\; d(x) \mapsto (-1)^{|x|} d(x), \tag{6.90} $$

turns an odd modular operad in Lin_S into one in Lin_M and vice versa. If not stated otherwise, all odd modular operads will be considered in Lin with the standard monoidal structure. For that reason we drop the subscript S.

Example 6.30 Recall [11, Definition 13] that an *anti-associative algebra* is a couple $A = (A, \star)$ consisting of a graded vector space A and a degree $+1$ operation $\star : A \otimes A \to A$ which is anti-associative, i.e.,

$$a \star (b \star c) + (-1)^{|a|}(a \star b) \star c = 0,$$

for all $a, b, c \in A$. An odd modular operad with step $s \neq 0$ such that $\mathscr{T}(S; g) \neq 0$ only if $\mathrm{card}(S) = 2$ and $g = 0$ is precisely an anti-associative algebra A with an involution $\tau : A \to A$ such that $\tau(ab) = \tau(b)\tau(a)$ for all $a, b \in A$, cf. Example 6.23.

Example 6.31 Let V be a graded vector space and $s \in V \otimes V$ a symmetric degree $+1$ tensor. The construction of the modular endomorphism operad given in Examples 6.6 and 6.28 translate verbatim, though the operations ${}_a\bullet_b$ and $\bullet_{uv}$ now have degree $+1$.

One must however be careful. While in the case of ordinary modular operads both constructions of the ${}_a\circ_b$ operation, i.e., the one via composition (6.15) and the one via composition (6.16) lead to the same results, now the results are different. The reason is that, while for $|s| = 0$ the above compositions are dual to each other, if $|s| = 1$ they are not, because the duality (2) acquires a nontrivial sign. One immediately sees that the resulting $f\, {}_a\bullet_b\, g$'s differ by $(-1)^{|f|+|g|}$.

Likewise, compositions (6.62) and (6.64) are in the $|s| = 1$ case not dual to each other and the resulting $\bullet_{uv}(f)$'s differ by $(-1)^{|f|}$. What happens is so surprising that we formulate it as a proposition; recall that Lin_S and Lin_M denote the two versions of the category Lin discussed in Remark 6.11.

Proposition 6.11 *For $|s| = 1$ compositions (6.16) and (6.64) lead to an odd modular operad in Lin_S, while compositions (6.15) and (6.62) to an odd modular operad in Lin_M.*

Proof. Let us show that the ${}_a\bullet_b$-operations defined by the odd version of (6.16) satisfy (6.86). Since $|\bar{s}| = |\bar{\bar{s}}|$, one must be cautious. We have

$$\begin{aligned} f\, {}_a\bullet_b(g\, {}_c\bullet_d\, h) &= \big(\mathbb{1}_{S_1} \otimes \bar{s} \otimes \mathbb{1}_{S_2 \sqcup S_3}\big)^{\#}\big(f \otimes (g\, {}_c\bullet_d\, h)\big) \\ &= \big(\mathbb{1}_{S_1} \otimes \bar{s} \otimes \mathbb{1}_{S_2 \sqcup S_3}\big)^{\#}\big(f \otimes (\mathbb{1}_{S_2 \sqcup \{b\}} \otimes \bar{\bar{s}} \otimes \mathbb{1}_{S_3})^{\#}(g \otimes h)\big) \\ &= (-1)^{|f|}\big(\mathbb{1}_{S_1} \otimes \bar{s} \otimes \mathbb{1}_{S_2 \sqcup S_3}\big)^{\#}\big(\mathbb{1}_{S_1 \sqcup \{a,b\} \sqcup S_2} \otimes \bar{\bar{s}} \otimes \mathbb{1}_{S_3}\big)^{\#}(f \otimes g \otimes h), \end{aligned}$$

while

$$\begin{aligned} (f\, {}_a\bullet_b\, g)\, {}_c\bullet_d\, h &= \big(\mathbb{1}_{S_1 \sqcup S_2} \otimes \bar{\bar{s}} \otimes \mathbb{1}_{S_3}\big)^{\#}\big((f\, {}_a\bullet_b\, g) \otimes h\big) \\ &= \big(\mathbb{1}_{S_1 \sqcup S_2} \otimes \bar{\bar{s}} \otimes \mathbb{1}_{S_3}\big)^{\#}\big((\mathbb{1}_{S_1} \otimes \bar{s} \otimes \mathbb{1}_{S_2 \sqcup \{c\}})^{\#}(f \otimes g) \otimes h\big) \\ &= \big(\mathbb{1}_{S_1 \sqcup S_2} \otimes \bar{\bar{s}} \otimes \mathbb{1}_{S_3}\big)^{\#}\big(\mathbb{1}_{S_1} \otimes \bar{s} \otimes \mathbb{1}_{S_2 \sqcup \{c,d\} \sqcup S_3}\big)^{\#}(f \otimes g \otimes h). \end{aligned}$$

To finish the proof of (6.86), we observe that

$$\begin{aligned}\big(\mathbb{1}_{S_1} \otimes \bar{s} \otimes \mathbb{1}_{S_2} \otimes \bar{\bar{s}} \otimes \mathbb{1}_{S_3}\big)^\# &= \big(\mathbb{1}_{S_1} \otimes \bar{s} \otimes \mathbb{1}_{S_2 \sqcup S_3}\big)^\# \big(\mathbb{1}_{S_1 \sqcup \{a,b\} \sqcup S_2} \otimes \bar{\bar{s}} \otimes \mathbb{1}_{S_3}\big)^\# \\ &= -\big(\mathbb{1}_{S_1 \sqcup S_2} \otimes \bar{\bar{s}} \otimes \mathbb{1}_{S_3}\big)^\# \big(\mathbb{1}_{S_1} \otimes \bar{s} \otimes \mathbb{1}_{S_2 \sqcup \{c,d\} \sqcup S_3}\big)^\#,\end{aligned}$$

the minus sign coming from commuting $\bar{s}$ over $\bar{\bar{s}}$.

Let us also verify that the ${}_a\bullet_b$-operations defined by the odd version of composition (6.15) satisfy (6.87). The related calculation is of course obtained from the above one by removing duals and inverting the order of compositions but, very crucially, *without* inserting Koszul signs. We obtain

$$\begin{aligned} f\ {}_a\bullet_b (g\ {}_c\bullet_d\ h) &= \big(f \otimes (g\ {}_c\bullet_d\ h)\big)\big(\mathbb{1}_{S_1} \otimes \bar{s} \otimes \mathbb{1}_{S_2 \sqcup S_3}\big) \\ &= \big(f \otimes (g \otimes h)(\mathbb{1}_{S_2 \sqcup \{b\}} \otimes \bar{\bar{s}} \otimes \mathbb{1}_{S_3})\big)\big(\mathbb{1}_{S_1} \otimes \bar{s} \otimes \mathbb{1}_{S_2 \sqcup S_3}\big) \\ &= (f \otimes g \otimes h)\big(\mathbb{1}_{S_1 \sqcup \{a,b\} \sqcup S_2} \otimes \bar{\bar{s}} \otimes \mathbb{1}_{S_3}\big)\big(\mathbb{1}_{S_1} \otimes \bar{s} \otimes \mathbb{1}_{S_2 \sqcup S_3}\big)\end{aligned}$$

on the one hand and

$$\begin{aligned} (f\ {}_a\bullet_b\ g)\ {}_c\bullet_d\ h &= \big((f\ {}_a\bullet_b\ g) \otimes h\big)\big(\mathbb{1}_{S_1 \sqcup S_2} \otimes \bar{\bar{s}} \otimes \mathbb{1}_{S_3}\big) \\ &= \big((f \otimes g)(\mathbb{1}_{S_1} \otimes \bar{s} \otimes \mathbb{1}_{S_2 \sqcup \{c\}}) \otimes h\big)\big(\mathbb{1}_{S_1 \sqcup S_2} \otimes \bar{\bar{s}} \otimes \mathbb{1}_{S_3}\big) \\ &= (-1)^{|h|}(f \otimes g \otimes h)\big(\mathbb{1}_{S_1} \otimes \bar{s} \otimes \mathbb{1}_{S_2 \sqcup \{c,d\} \sqcup S_3}\big)\big(\mathbb{1}_{S_1 \sqcup S_2} \otimes \bar{\bar{s}} \otimes \mathbb{1}_{S_3}\big)\end{aligned}$$

on the other hand. Axiom (6.87) now follows from the equality

$$\begin{aligned}\big(\mathbb{1}_{S_1} \otimes \bar{s} \otimes \mathbb{1}_{S_2} \otimes \bar{\bar{s}} \otimes \mathbb{1}_{S_3}\big) &= \big(\mathbb{1}_{S_1} \otimes \bar{s} \otimes \mathbb{1}_{S_2 \sqcup \{c,d\} \sqcup S_3}\big)\big(\mathbb{1}_{S_1 \sqcup S_2} \otimes \bar{\bar{s}} \otimes \mathbb{1}_{S_3}\big) \\ &= -\big(\mathbb{1}_{S_1 \sqcup \{a,b\} \sqcup S_2} \otimes \bar{\bar{s}} \otimes \mathbb{1}_{S_3}\big)\big(\mathbb{1}_{S_1} \otimes \bar{s} \otimes \mathbb{1}_{S_2 \sqcup S_3}\big).\end{aligned}$$

Notice that the sign difference between the results of the above two computations is $(-1)^{|x|}$ versus $(-1)^{|z|}$ as it should be. The verification of axioms (6.88) resp. (6.89) is similar.

We will call the family (6.61) with the ${}_a\bullet_b$- and $\bullet_{uv}$-operations defined via compositions (6.16) and (6.64) with $|s| = 1$ the *odd* modular endomorphism operad and denote it $\mathcal{E}nd_V$; whether $\mathcal{E}nd_V$ means the odd modular endomorphism operad or the ordinary one (with $|s| = 0$) will always be clear from the context. For a degree $+1$ bilinear form $B : V \otimes V \to \Bbbk$ we also have the odd modular version of the operad $\mathcal{D}ne_V$ of Example 6.7. We leave the details to the reader.

Remark 6.12 Odd modular operads have their skeletal versions. For $\mathscr{T}$ as in (6.77), $n \geq 0$ and $g \in \mathbb{A}$, denote $\mathscr{T}(n; g) := \mathscr{T}([n]; g)$. The degree $+1$ operations

$$ {}_i\overline{\bullet}_j : \mathscr{T}(m+1; g_1) \otimes \mathscr{T}(n+1; g_2) \to \mathscr{T}(m+n; g_1+g_2) $$

and

$$ \overline{\bullet}_{ij} : \mathscr{T}(n+2; g+s) \to \mathscr{T}(n; g),\ 1 \leq i, j \leq n+2,\ g \in \mathbb{A} $$

are defined by obvious formulas analogous to (6.8) resp. (6.70). The axioms for these operations are the same as the skeletal axioms for modular operads, only the axioms corresponding to (6.80)–(6.83) acquire the minus sign.

Example 6.32 The skeletal version of the odd endomorphism operad from Example 6.31 is described as follows. One has $\mathscr{E}nd_V(n; g) \cong Lin(V^{\otimes n}, \Bbbk)$ for $n \geq 0$, $g \in \mathbb{A}$ as in (6.72), with the skeletal ${}_i\overline{\bullet}_j$-operations defined by formula (6.25), but this time with

$$ \begin{aligned} \kappa = {} & |g|(|f|+1) + |s_i''| + |s_i'|(|v_{i+n}| + \cdots + |v_{m+n}|) \\ & + |s_i''|(|v_{n+i-j+1}| + \cdots + |v_{i+n-1}|). \end{aligned} $$

The skeletal contractions $\overline{\bullet}_{ij}$ are given by formula (6.73) with

$$ \kappa = |f| + (|v_1| + \cdots + |v_{i-1}|) + |s_i''|(|v_i| + \cdots + |v_{j-2}|). $$

We leave the related straightforward verification to the reader.

Odd endomorphism operads are necessary for the definition of algebras over odd modular operads.

Definition 6.25 Let V be a graded vector space and $s \in V \otimes V$ a symmetric degree $+1$ tensor. An *algebra* over an odd modular operad $\mathscr{T}$, or a $\mathscr{T}$*-algebra*, is a morphism $\alpha : \mathscr{T} \to \mathscr{E}nd_V$ of odd modular operads.

The main source of examples of algebras over odd modular operads will be provided by algebras over the Feynman transform of a modular operads introduced in Sect. 7.2.

Definition 6.26 For a finite set X, let $\uparrow \Bbbk^X$ be the free $\Bbbk$-module with basis X, considered as a dg-vector space concentrated in degree 1. Define the one-dimensional dg-vector space concentrated in degree card(X):

$$ \det(X) = \det(\Bbbk^X) := \wedge^{|X|}(\uparrow \Bbbk^X), $$

where $\wedge^{|X|}(-)$ denotes the $|X|$th exterior (Grassmann) power. Since $\Bbbk^X$ is the space of $\Bbbk$-valued functions on X, $\det(X)$ is "contravariant in X."

Free odd modular operads are constructed as the ordinary ones, with the twisting build into the construction via the determinant of the set of edges of the underlying graphs. Explicitly, for a labeled graph $\boldsymbol{\Gamma}$ and a modular module E we take the odd analog

$$\widetilde{E}(\boldsymbol{\Gamma}) := \det\big(Edg(\Gamma)\big) \otimes \bigotimes_{v \in Vert(\Gamma)} E\big(Leg(v); \ell(v)\big)$$

of (6.33). Each isomorphism $\phi : \boldsymbol{\Gamma}_0 \xrightarrow{\cong} \boldsymbol{\Gamma}_1$ of labeled graphs again induces an isomorphism

$$\widetilde{E}(\phi) : \widetilde{E}(\boldsymbol{\Gamma}_0) \xrightarrow{\cong} \widetilde{E}(\boldsymbol{\Gamma}_1).$$

For a finite set S and a genus $g \in \mathbb{A}$ we define

$$\widetilde{\mathbb{F}}(E)(S; g) := \frac{\bigoplus_G \widetilde{E}(\boldsymbol{\Gamma})}{\sim}$$

with the sum taken over all labeled graphs $\boldsymbol{\Gamma} = (\Gamma, \ell)$ with $Leg(\Gamma) = S$ and $g(\boldsymbol{\Gamma}) = g$. As before, the relation $\sim$ identifies $x \in \widetilde{E}(\boldsymbol{\Gamma}_0)$ with its image $\widetilde{E}(\phi)(x) \in \widetilde{E}(\boldsymbol{\Gamma}_1)$ for any isomorphism $\phi : \boldsymbol{\Gamma}_0 \to \boldsymbol{\Gamma}_1$ that induces the identity of the set of the legs.

Proposition 6.12 *The modular module $\widetilde{\mathbb{F}}(E) = \big\{\widetilde{\mathbb{F}}(E)(S; g) \mid S \in \mathtt{Cor},\ g \in \mathbb{A}\big\}$ is an odd modular operad.*

Proof. Isomorphisms of finite sets act on $\widetilde{\mathbb{F}}(E)$ by relabeling the legs of the underlying graphs. With this action, $\widetilde{\mathbb{F}}(E)$ is a modular module. For the grafting $\boldsymbol{\Gamma}_1\, {}_a\circ_b\, \boldsymbol{\Gamma}_2$ of labeled graphs in (6.65) one clearly has

$$\uparrow\big(\widetilde{E}(\boldsymbol{\Gamma}') \otimes \widetilde{E}(\boldsymbol{\Gamma}'')\big) \cong \det\big(\{e\}\big) \otimes \widetilde{E}(\boldsymbol{\Gamma}_1) \otimes \widetilde{E}(\boldsymbol{\Gamma}_2) \cong \widetilde{E}(\boldsymbol{\Gamma}_1\, {}_a\circ_b\, \boldsymbol{\Gamma}_2)$$

with $e := \{a, b\}$ the newly created edge of the underlying graph $(\Gamma_1\, {}_a\circ_b\, \Gamma_2)$. One then has the induced composed morphism of the quotients

$$\widetilde{\mathbb{F}}(E)\big(S_1 \sqcup \{a\}, g_1\big) \otimes \widetilde{\mathbb{F}}(E)\big(S_2 \sqcup \{b\}, g_2\big) \xrightarrow{\uparrow}$$
$$\uparrow\big(\widetilde{\mathbb{F}}(E)\big(S_1 \sqcup \{a\}, g_1\big) \otimes \widetilde{\mathbb{F}}(E)\big(S_2 \sqcup \{b\}, g_2\big)\big) \xrightarrow{\cong} \widetilde{\mathbb{F}}(E)(S_1 \sqcup S_2, g_1 + g_2)$$

which we take as a definition of the compositions.

The contractions are constructed similarly, this time using the isomorphism

$$\uparrow \widetilde{E}(\boldsymbol{\Gamma}) \cong \det\big(\{f\}\big) \otimes \widetilde{E}(\boldsymbol{\Gamma}) \cong \widetilde{E}(\circ_{uv}\boldsymbol{\Gamma}; g),$$

where $\circ_{uv}\boldsymbol{\Gamma}$ is as in Definition 6.22, and $f := \{u, v\}$. The contraction is then the composed map

$$\widetilde{\mathbb{F}}(E)\big(S \sqcup \{u, v\}; g\big) \xrightarrow{\uparrow} \uparrow \big(\widetilde{\mathbb{F}}(E)\big(S \sqcup \{u, v\}; g\big)\big) \xrightarrow{\cong} \widetilde{\mathbb{F}}(E)(S; g + s).$$

Thanks to the presence of the suspensions, the operations ${}_a\circ_b$ and $\circ_{uv}$ constructed above have degree $+1$ as required. The axioms of odd modular operads can be verified directly.

Proposition 6.13 *The odd modular operad $\widetilde{\mathbb{F}}(E)$ is the free odd modular operad on the modular module E.*

Proof. The operad $\widetilde{\mathbb{F}}(E)$ is the triple (monad) $\mathbb{M}_{\mathfrak{D}}$ with $\mathfrak{D}$ the dualizing cocycle $\mathfrak{K}$, evaluated at the modular module E, see Theorem 5.47 and Example 5.52 of [12] for $\mathbb{M}_{\mathfrak{D}}$ and $\mathfrak{K}$, respectively.

As we already know, removing the contractions $\circ_{uv}$ and the genus grading from the definition of (ordinary) modular leads to cyclic operads. A natural question is what happens if we do the same in the definition of odd modular operads. We obtain:

Definition 6.27 An *odd cyclic operad* $\mathscr{P}$ is a cyclic module

$$\mathscr{P} = \big\{\mathscr{P}(C) \in \mathtt{Chain} \mid C \in \mathtt{Cor}\big\}$$

together with degree $+1$ morphisms

$${}_a\circ_b : \mathscr{P}\big(C_1 \sqcup \{a\}\big) \otimes \mathscr{P}\big(C_2 \sqcup \{b\}\big) \to \mathscr{P}(C_1 \sqcup C_2)$$

defined for arbitrary disjoint finite sets C_1, C_2 and symbols a, b. These data satisfy verbatim analogs of axioms (i)–(iv) of Definition 6.1, except that equality (6.3) now involves the minus sign, i.e., it reads

$${}_a\circ_b(\mathbb{1} \otimes {}_c\circ_d) = -\, {}_c\circ_d({}_a\circ_b \otimes \mathbb{1}). \tag{6.91}$$

Odd cyclic operads are however not interesting per se because they are desuspensions of ordinary cyclic operads. More precisely, for a cyclic operad $\mathscr{P}$ as in Definition 6.1 consider the cyclic module

$$\downarrow \mathscr{P} = \big\{ \downarrow \mathscr{P}(C) \in \mathtt{Chain} \mid C \in \mathtt{Cor}\big\},$$

where $\downarrow \mathscr{P}(C)$ is the desuspension of the dg-vector space $\mathscr{P}(C)$, i.e., $\mathscr{P}(C)$ with degrees shifted down by one. We define the composition operations of $\downarrow \mathscr{P}$ using the ones of $\mathscr{P}$ as the composition

$$\downarrow \mathscr{P}\big(C_1 \sqcup \{a\}\big)\otimes \downarrow \mathscr{P}\big(C_2 \sqcup \{b\}\big) \xrightarrow{\uparrow\otimes\uparrow}$$
$$\mathscr{P}\big(C_1 \sqcup \{a\}\big) \otimes \mathscr{P}\big(C_2 \sqcup \{b\}\big) \xrightarrow{{}_a\circ_b} \mathscr{P}(C_1 \sqcup C_2) \xrightarrow{\downarrow} \downarrow \mathscr{P}(C_1 \sqcup C_2).$$

It is easy to verify that the above construction is an odd modular operad. The following claim is obvious.

Proposition 6.14 *The correspondence $\mathscr{P} \longmapsto \downarrow \mathscr{P}$ is an equivalence between the category of cyclic operads and the category of odd cyclic operads.*

It is however not true that the desuspension of a modular operad is an odd modular operad. For instance, the induced contractions on the desuspension would have degree 0 not $+1$ as required.

References

1. Batanin, M.: Monoidal globular categories as a natural environment for the theory of weak n-categories. Adv. Math. **136**(1), 39–103 (1998). http://dx.doi.org/10.1006/aima.1998.1724
2. Getzler, E., Kapranov, M.: Cyclic operads and cyclic homology. In: Yau, S.T. (ed.) Geometry, Topology and Physics for Raoul Bott. Conference Proceedings. Lecture Notes in Geometry and Topology, vol. 4, pp. 167–201. International Press, Boston (1995)
3. Getzler, E., Kapranov, M.: Modular operads. Compos. Math. **110**(1), 65–126 (1998)
4. Ginzburg, V., Kapranov, M.: Koszul duality for operads. Duke Math. J. **76**(1), 203–272 (1994)
5. Hartshorne, R.: Algebraic Geometry. Graduate Texts in Mathematics, vol. 52. Springer, New York (1977)
6. Mac Lane, S.: Categories for the Working Mathematician. Graduate Texts in Mathematics, vol. 5, 2nd edn. Springer, New York (1998)
7. Markl, M.: Models for operads. Commun. Algebra **24**(4), 1471–1500 (1996). http://dx.doi.org/10.1080/00927879608825647
8. Markl, M.: Loop homotopy algebras in closed string field theory. Commun. Math. Phys. **221**(2), 367–384 (2001). http://dx.doi.org/10.1007/PL00005575
9. Markl, M.: Operads and PROPs. In: Handbook of Algebra, vol. 5, pp. 87–140. Elsevier/North-Holland, Amsterdam (2008). http://dx.doi.org/10.1016/S1570-7954(07)05002-4
10. Markl, M.: Modular envelopes, OSFT and nonsymmetric (non-Σ) modular operads. J. Noncommut. Geom. **10**(2), 775–809 (2016). http://dx.doi.org/10.4171/JNCG/248
11. Markl, M., Remm, E.: (Non-)Koszulness of operads for n-ary algebras, galgalim and other curiosities. J. Homotopy Relat. Struct. **10**, 939–269 (2015)
12. Markl, M., Shnider, S., Stasheff, J.: Operads in Algebra, Topology and Physics. Mathematical Surveys and Monographs, vol. 96. American Mathematical Society, Providence (2002)

7 Feynman Transform of a Modular Operad

The aim of this chapter is to recall an analog of the bar construction for modular operads, called in this context the Feynman transform and introduced in [2], see also [3, Section II.5.3].

7.1 Modules and Derivations

We define modules over odd modular operads and derivations with values in these modules. We then introduce trivial extensions of odd operads by these modules and prove that derivations can be encoded by morphisms of these extensions. This will imply that a derivation whose source is a free odd modular operad is uniquely determined by its restriction to the modular module of generators. This scheme is standard, but some care is needed to get signs right, as the operations of odd operads and their modules have degree $+1$.

Definition 7.1 Let $\mathscr{T}$ be an odd modular operad with compositions ${}_a\bullet_b$ and contractions $\bullet_{uv}$ as in Definition 6.24. A *$\mathscr{T}$-module*, or a *module over* $\mathscr{T}$ is a modular module

$$\mathsf{M} = \big\{\mathsf{M}(S; g) \in \mathtt{Chain};\ (S; g) \in \mathtt{Cor} \times \mathbb{A}\big\}$$

together with degree +1 morphisms

$${}_a\overset{L}{\bullet}_b : \mathscr{T}\big(S_1 \sqcup \{a\}; g_1\big) \otimes \mathsf{M}\big(S_2 \sqcup \{b\}; g_2\big) \to \mathsf{M}(S_1 \sqcup S_2; g_1 + g_2)$$

defined for arbitrary disjoint finite sets S_1, S_2, symbols a, b, and genera $g_1, g_2 \in \mathbb{A}$, and degree +1 maps

$$\blacklozenge_{uv} = \blacklozenge_{vu} : \mathsf{M}\big(S \sqcup \{u, v\}; g\big) \to \mathsf{M}(S; g + s)$$

© Springer Nature Switzerland AG 2020

M. Doubek et al., *Algebraic Structure of String Field Theory*, Lecture Notes in Physics 973, https://doi.org/10.1007/978-3-030-53056-3_7

given for any finite set S, genus $g \in \mathbb{A}$, and symbols u, v. These data satisfy the following axioms.

(i) For arbitrary isomorphisms $\rho : S_1 \sqcup \{a\} \to T_1$, $\sigma : S_2 \sqcup \{b\} \to T_2$ of finite sets and genera $g_1, g_2 \in \mathbb{A}$, one has the equality

$$\mathsf{M}\big(\rho|_{S_1} \sqcup \sigma|_{S_2}\big)\, {}_a\overset{L}{\bullet}_b = {}_{\rho(a)}\overset{L}{\bullet}_{\sigma(b)} \, \big(\mathscr{T}(\rho) \otimes \mathsf{M}(\sigma)\big)$$

of maps

$$\mathscr{T}\big(S_1 \sqcup \{a\}; g_1\big) \otimes \mathsf{M}\big(S_2 \sqcup \{b\}; g_2\big) \to \mathsf{M}\big(T_1 \sqcup T_2 \setminus \{\rho(a), \sigma(b)\}; g_1 + g_2\big).$$

(ii) For any isomorphism $\rho : S \sqcup \{u, v\} \to T$ of finite sets and a genus $g \in \mathbb{A}$, one has the equality

$$\mathsf{M}\big(\rho|_S\big)\, \blacklozenge_{uv} = \blacklozenge_{\rho(u)\rho(v)} \mathsf{M}(\rho)$$

of maps $\mathsf{M}\big(S \sqcup \{u, v\}; g\big) \to \mathsf{M}\big(T \setminus \{\rho(u), \rho(v)\}; g + s\big)$.

(iii) For mutually disjoint sets S_1, S_2, S_3, symbols a, b, c, d, and genera $g_1, g_2, g_3 \in \mathbb{A}$, one has the equality

$$ {}_a\overset{L}{\bullet}_b(\mathbb{1} \otimes {}_c\overset{L}{\bullet}_d) = -\, {}_c\overset{L}{\bullet}_d(\, {}_a\bullet_b \otimes \mathbb{1}) \tag{7.1}$$

of maps from $\mathscr{T}\big(S_1 \sqcup \{a\}; g_1\big) \otimes \mathscr{T}\big(S_2 \sqcup \{b, c\}; g_2\big) \otimes \mathsf{M}\big(S_3 \sqcup \{d\}; g_3\big)$ to the space $\mathsf{M}\big(S_1 \sqcup S_2 \sqcup S_3; g_1 + g_2 + g_3\big)$.

(iv) For $S_1, S_2, S_3, a, b, c, d$ and $g_1, g_2, g_3 \in \mathbb{A}$ as in (iii), one has the equality

$$ {}_a\overset{L}{\bullet}_b\, (\mathbb{1} \otimes {}_d\overset{L}{\bullet}_c) = -\, {}_d\overset{L}{\bullet}_c\, (\mathbb{1} \otimes {}_a\overset{L}{\bullet}_b)(\tau \otimes \mathbb{1})$$

of maps from $\mathscr{T}\big(S_1 \sqcup \{a\}; g_1\big) \otimes \mathscr{T}\big(S_2 \sqcup \{d\}; g_2\big) \otimes \mathsf{M}\big(S_3 \sqcup \{b, c\}; g_3\big)$ to the space $\mathsf{M}\big(S_1 \sqcup S_2 \sqcup S_3; g_1 + g_2 + g_3\big)$.

(v) For a finite set S, symbols a, b, c, d and a genus $g \in \mathbb{A}$ one has the equality

$$\blacklozenge_{ab} \blacklozenge_{cd} = -\, \blacklozenge_{cd} \blacklozenge_{ab}$$

of maps $\mathsf{M}\big(S \sqcup \{a, b, c, d\}; g\big) \to \mathsf{M}(S; g + 2s)$.

(vi) For finite sets S_1, S_2, symbols a, b, c, d and genera $g_1, g_2 \in \mathbb{A}$, one has the equality

$$\blacklozenge_{ab}\, {}_c\overset{L}{\bullet}_d = -\, \blacklozenge_{cd}\, {}_a\overset{L}{\bullet}_b$$

of maps $\mathscr{T}\big(S_1 \sqcup \{a, c\}; g_1\big) \otimes \mathsf{M}\big(S_2 \sqcup \{b, d\}; g_2\big) \to \mathsf{M}(S_1 \sqcup S_2; g_1 + g_2 + s)$.

(vii) For finite sets S_1, S_2, symbols a, b, u, v, and genera $g_1, g_2 \in \mathbb{A}$, one has the equality

$$ {}_a\overset{L}{\bullet}_b\, (\bullet_{uv} \otimes \mathbb{1}) = -\, \blacklozenge_{uv}\ {}_a\overset{L}{\bullet}_b \tag{7.2}$$

of maps $\mathscr{T}\big(S_1 \sqcup \{a, u, v\}; g_1\big) \otimes \mathsf{M}\big(S_2 \sqcup \{b\}; g_2\big) \to \mathsf{M}(S_1 \sqcup S_2; g_1 + g_2 + s)$.

(viii) For finite sets S_1, S_2, symbols a, b, u, v, and genera $g_1, g_2 \in \mathbb{A}$, one has the equality

$$ {}_a\overset{L}{\bullet}_b\, (\mathbb{1} \otimes \blacklozenge_{uv}) = -\, \blacklozenge_{uv}\ {}_a\overset{L}{\bullet}_b $$

of maps $\mathscr{T}\big(S_1 \sqcup \{a\}; g_1\big) \otimes \mathsf{M}\big(S_2 \sqcup \{b, u, v\}; g_2\big) \to \mathsf{M}(S_1 \sqcup S_2; g_1 + g_2 + s)$.

Remark 7.1 The superscript "L" of ${}_a\overset{L}{\bullet}_b$ reminds us that we apply an element of $\mathscr{T}$ from the left to an element of M. It will be convenient to define also the auxiliary right composition

$$ {}_a\overset{R}{\bullet}_b : \mathsf{M}(S_1 \sqcup \{a\}; g_1) \otimes \mathscr{T}(S_2 \sqcup \{b\}; g_2) \to \mathsf{M}(S_1 \sqcup S_2; g_1 + g_2) $$

as ${}_a\overset{R}{\bullet}_b := {}_b\overset{L}{\bullet}_a\, \tau$, i.e. on elements as

$$ x\ {}_a\overset{R}{\bullet}_b\ y := (-1)^{|x||y|}\, y\ {}_b\overset{L}{\bullet}_a\ x $$

for $x \in \mathsf{M}\big(S_1 \sqcup \{a\}; g_1\big)$ and $y \in \mathscr{T}\big(S_2 \sqcup \{b\}; g_2\big)$. One easily proves the following properties of this operation.

(i) For arbitrary isomorphisms $\rho : S_1 \sqcup \{a\} \to T_1$ and $\sigma : S_2 \sqcup \{b\} \to T_2$ of finite sets and genera $g_1, g_2 \in \mathbb{A}$, one has the equality

$$ \mathsf{M}(\rho|_{S_1} \sqcup \sigma|_{S_2})\ {}_a\overset{R}{\bullet}_b = {}_{\rho(a)}\overset{R}{\bullet}_{\sigma(b)}\ (\mathsf{M}(\rho) \otimes \mathscr{T}(\sigma)) $$

of maps

$$ \mathsf{M}\big(S_1 \sqcup \{a\}; g_1\big) \otimes \mathscr{T}\big(S_2 \sqcup \{b\}; g_2\big) \to \mathsf{M}\big(T_1 \sqcup T_2 \setminus \{\rho(a), \sigma(b)\}; g_1 + g_2\big). $$

(ii) For finite sets S_1, S_2, symbols a, b, c, d and genera $g_1, g_2 \in \mathbb{A}$, one has the equality

$$ \blacklozenge_{ab}\ {}_c\overset{R}{\bullet}_d = -\, \blacklozenge_{cd}\ {}_a\overset{R}{\bullet}_b $$

of maps $\mathsf{M}\big(S_1 \sqcup \{a, c\}; g_1\big) \otimes \mathscr{T}\big(S_2 \sqcup \{b, d\}; g_2\big) \to \mathsf{M}(S_1 \sqcup S_2; g_1 + g_2 + s)$.

(iii) For finite sets S_1, S_2, symbols a, b, u, v, and genera $g_1, g_2 \in \mathbb{A}$, one has the equality

$$ {}_a\overset{R}{\bullet}_b\,(\blacklozenge_{uv} \otimes \mathbb{1}) = -\,\blacklozenge_{uv}\;{}_a\overset{R}{\bullet}_b $$

of maps $\mathsf{M}(S_1 \sqcup \{a, u, v\}; g_1) \otimes \mathscr{T}(S_2 \sqcup \{b\}; g_2) \to \mathsf{M}(S_1 \sqcup S_2; g_1 + g_2 + s)$.

(iv) For finite sets S_1, S_2, symbols a, b, u, v, and genera $g_1, g_2 \in \mathbb{A}$, one has the equality

$$ {}_a\overset{R}{\bullet}_b\,(\mathbb{1} \otimes \bullet_{uv}) = -\blacklozenge_{uv}\;{}_a\overset{R}{\bullet}_b $$

of maps $\mathsf{M}(S_1 \sqcup \{a\}; g_1) \otimes \mathscr{T}(S_2 \sqcup \{b, u, v\}; g_2) \to \mathsf{M}(S_1 \sqcup S_2; g_1 + g_2 + s)$.

(v) For mutually disjoint sets S_1, S_2, S_3, symbols a, b, c, d and genera $g_1, g_2, g_3 \in \mathbb{A}$, one has the equality

$$ {}_a\overset{L}{\bullet}_b\,(\mathbb{1} \otimes {}_c\overset{R}{\bullet}_d) = -\,{}_c\overset{R}{\bullet}_d\,({}_a\overset{L}{\bullet}_b \otimes \mathbb{1}) \tag{7.3} $$

of maps from $\mathscr{T}(S_1 \sqcup \{a\}; g_1) \otimes \mathsf{M}(S_2 \sqcup \{b, c\}; g_2) \otimes \mathscr{T}(S_3 \sqcup \{d\}; g_3)$ to the space $\mathsf{M}(S_1 \sqcup S_2 \sqcup S_3; g_1+g_2+g_3)$.

(vi) For mutually disjoint sets S_1, S_2, S_3, symbols a, b, c, d and genera $g_1, g_2, g_3 \in \mathbb{A}$, one has the equality

$$ {}_a\overset{R}{\bullet}_b\,(\mathbb{1} \otimes {}_c\bullet_d) = -\,{}_c\overset{R}{\bullet}_d\,({}_a\overset{R}{\bullet}_b \otimes \mathbb{1}) \tag{7.4} $$

of maps from $\mathsf{M}(S_1 \sqcup \{a\}; g_1) \otimes \mathscr{T}(S_2 \sqcup \{b, c\}; g_2) \otimes \mathscr{T}(S_3 \sqcup \{d\}; g_3)$ to the space $\mathsf{M}(S_1 \sqcup S_2 \sqcup S_3; g_1+g_2+g_3)$.

Since the pasting schemes for modular operads and their modules are abstract graphs, there is no concept of "left" and "right" compositions so, unlike, e.g. left versus right modules over associative algebras, ${}_a\overset{L}{\bullet}_b$ and ${}_a\overset{R}{\bullet}_b$ are materializations of the same operation which differ only by the way they are written on paper.

Example 7.1 Any odd modular operad $\mathscr{T}$ with the structure operations ${}_a\bullet_b$ and $\bullet_{uv}$ is a module over itself, with

$$ {}_a\overset{L}{\bullet}_b := {}_a\bullet_b \ \text{ and } \ \blacklozenge_{uv} := \bullet_{uv}. $$

All axioms of Definition 7.1 clearly follow from Definition 6.24 of an odd operad. Slightly less straightforward is only (iv) obtained by applying $(\mathbb{1} \otimes \tau)$ from the right on (6.80), and (vii) obtained by applying τ from the right on (6.83).

Definition 7.2 Let M be a module over an odd operad $\mathscr{T}$. A degree k *derivation* $\theta : \mathscr{T} \to \mathsf{M}$ is a degree k morphism

$$\theta = \big\{\theta(S; g) : \mathscr{T}(S; g) \to \mathsf{M}(S; g);\ (S; g) \in \mathtt{Cor} \times \mathbb{A}\big\}$$

of modular modules[1] satisfying, for finite S_1, S_2, symbols a, b, and genera $g_1, g_2 \in \mathbb{A}$ the equality

$$\theta\ {}_a\bullet_b = (-1)^k\big({}_a\overset{R}{\bullet}_b(\theta \otimes \mathbb{1}) + {}_a\overset{L}{\bullet}_b(\mathbb{1} \otimes \theta)\big), \tag{7.5}$$

of maps $\mathscr{T}\big(S_1 \sqcup \{a\}; g_1\big) \otimes \mathscr{T}\big(S_2 \sqcup \{b\}; g_2\big) \to \mathsf{M}(S_1 \sqcup S_2; g_1 + g_2)$, and for a finite set S, symbols u, v, and a genus $g \in \mathbb{A}$ the equality

$$\theta \bullet_{ab} = (-1)^k \blacklozenge_{ab}\theta \tag{7.6}$$

of maps $\mathscr{T}\big(S \sqcup \{u, v\}; g\big) \to \mathscr{T}(S; g + s)$.

One can of course replace ${}_a\overset{R}{\bullet}_b(\theta\otimes\mathbb{1})$ in (7.5) by ${}_b\overset{L}{\bullet}_a\,\tau(\theta\otimes\mathbb{1}) = {}_b\overset{L}{\bullet}_a(\mathbb{1}\otimes\theta)\tau$, but using ${}_a\overset{R}{\bullet}_b$ makes the analogy between (7.5) and, e.g. the standard Leibniz property of derivations of associative algebras manifest. The auxiliary ${}_a\overset{R}{\bullet}_b$ will also simplify some formulas in the proofs that follow.

Example 7.2 The case when $\mathsf{M} = \mathscr{T}$, considered as a module over itself as in Example 7.1, is particularly important. With this convention, an example of a degree $+1$ derivation is the internal differential $d_{\mathscr{T}} : \mathscr{T} \to \mathscr{T}$ of an odd modular operad.

The following lemma in which an odd operad $\mathscr{T}$ is considered as a module over itself will be used in the proof of Theorem 7.2.

Lemma 7.1 *Let $\alpha, \beta : \mathscr{T} \to \mathscr{T}$ be derivations of an odd modular operad $\mathscr{T}$ such that α is of degree k and β of degree l. Then the linear map*

$$[\alpha, \beta] := \alpha\beta - (-1)^{kl}\beta\alpha : \mathscr{T} \to \mathscr{T}$$

is a degree $k{+}l$ derivation. In particular, if $k = l = 1$, then both α^2, β^2 and $\alpha\beta+\beta\alpha$ are degree 2 derivations.

[1] See Remark 6.10.

Proof. For homogeneous $x \in \mathscr{T}\big(S_1 \sqcup \{a\}; g_1\big)$ and $y \in \mathscr{T}\big(S_2 \sqcup \{a\}; g_2\big)$, one has

$$\begin{aligned}\alpha\beta(x\,{}_a\bullet_b\,y) &= (-1)^l\alpha\big(\beta(x)\,{}_a\bullet_b\,y + (-1)^{l|x|}x\,{}_a\bullet_b\,\beta y\big)\\ &= (-1)^{k+l}\Big(\alpha\beta(x)\,{}_a\bullet_b\,y + (-1)^{(l+|x|)k}\,\beta(x)\,{}_a\bullet_b\,\alpha(y)\\ &\quad + (-1)^{l|x|}\alpha(x)\,{}_a\bullet_b\,\beta(y) + (-1)^{(k+l)|x|}x\,{}_a\bullet_b\,\alpha\beta(y)\Big).\end{aligned}$$

Likewise

$$\begin{aligned}-(-1)^{kl}\,\beta\alpha(x\,{}_a\bullet_b\,y) &= -(-1)^{kl}(-1)^k\beta\big(\alpha(x)\,{}_a\bullet_b\,y + (-1)^{k|x|}x\,{}_a\bullet_b\,\alpha y\big)\\ &= (-1)^{k+l}\Big(-(-1)^{kl}\,\beta\alpha(x)\,{}_a\bullet_b\,y - (-1)^{l|x|}\alpha(x)\,{}_a\bullet_b\,\beta(y)\\ &\quad - (-1)^{(k+l)|x|}\beta(x)\,{}_a\bullet_b\,\alpha(y) - (-1)^{kl+(k+l)|x|}x\,{}_a\bullet_b\,\beta\alpha(y)\Big).\end{aligned}$$

Summing the above two equations we get that

$$[\alpha,\beta](x\,{}_a\bullet_b\,y) = (-1)^{(k+l)}\big([\alpha,\beta](x)\,{}_a\bullet_b\,y + (-1)^{(k+l)|x|}x\,{}_a\bullet_b[\alpha,\beta](y)\big),$$

which is (7.5) for $\theta = [\alpha,\beta]$ evaluated at $x \otimes y$. In the same vein, one obtains for $z \in \mathscr{T}\big(S \sqcup \{u,v\}\big)$ that

$$\alpha\beta(\bullet_{uv}z) = (-1)^l\alpha\,\bullet_{uv}\,\beta(z) = (-1)^{k+l}\,\bullet_{uv}\,\alpha\beta(z)$$

while

$$-(-1)^{kl}\beta\alpha(\bullet_{uv}z) = -(-1)^{kl}(-1)^k\beta\,\bullet_{uv}\,\alpha(z) = (-1)^{k+l}\big(-(-1)^{kl}\,\bullet_{uv}\,\beta\alpha(z)\big)$$

therefore

$$[\alpha,\beta](\bullet_{uv}z) = (-1)^{k+l}\,\bullet_{uv}\,\big([\alpha,\beta](z)\big),$$

which is (7.6) for $\theta = [\alpha,\beta]$ evaluated at z. To prove the last sentence of the lemma, one needs to observe that if $\deg(\alpha) = \deg(\beta) = 1$, then $\alpha^2 = 2[\alpha,\alpha]$, $\beta^2 = 2[\beta,\beta]$ and $\alpha\beta + \beta\alpha = [\alpha,\beta]$.

Definition 7.3 The *trivial extension* of an odd modular operad $\mathscr{T}$ by a $\mathscr{T}$-module M is an odd modular operad $\mathscr{T} \oplus \mathsf{M}$ whose underlying modular module is defined by

$$(\mathscr{T} \oplus \mathsf{M})(S;g) := \mathscr{T}(S;g) \oplus \mathsf{M}(S;g),\ \ (S;g) \in \mathtt{Cor} \times \mathbb{A},$$

with the obvious diagonal action of isomorphisms of finite sets. The structure operations (6.78) are given by

$$(t_1, m_1)\,{}_a\bullet_b(t_2, m_2) := (t_1\,{}_a\bullet_b\,t_2,\, t_1\,{}_a\overset{L}{\bullet}_b\,m_2 + m_1\,{}_a\overset{R}{\bullet}_b\,t_2),$$

where

$$t_1 \in \mathscr{T}\big(S_1 \sqcup \{a\}; g_1\big),\ \ t_2 \in \mathscr{T}\big(S_2 \sqcup \{b\}; g_2\big)$$

and

$$m_1 \in \mathsf{M}\big(S_1 \sqcup \{a\}; g_1\big),\ \ m_2 \in \mathsf{M}\big(S_2 \sqcup \{b\}; g_2\big).$$

The contractions (6.79) are defined diagonally,

$$\bullet_{uv}(t, m) := (\bullet_{uv}t, \blacklozenge_{uv}m).$$

for $t \in \mathscr{T}\big(S \sqcup \{u, v\}; g\big)$ and $m \in \mathsf{M}\big(S \sqcup \{u, v\}; g\big)$.

Lemma 7.2 *The modular module $\mathscr{T} \oplus \mathsf{M}$ with the operations defined above is an odd modular operad.*

Proof. The axioms of modular operads can be verified directly. Let us, for instance, verify axiom (6.80) of Definition 6.24. For

$$t_1 \in \mathscr{T}\big(S_1 \sqcup \{a\}; g_1\big),\ \ t_2 \in \mathscr{T}\big(S_2 \sqcup \{b, c\}; g_2\big),\ \ t_3 \in \mathscr{T}\big(S_3 \sqcup \{d\}; g_3\big)$$

and

$$m_1 \in \mathsf{M}\big(S_1 \sqcup \{a\}; g_1\big),\ \ m_2 \in \mathsf{M}\big(S_2 \sqcup \{b, c\}; g_2\big),\ \ m_3 \in \mathsf{M}\big(S_3 \sqcup \{d\}; g_3\big)$$

one has

$$\begin{aligned}
&{}_a\bullet_b(\mathbb{1} \otimes {}_c\bullet_d)\big((t_1, m_1) \otimes (t_2, m_2) \otimes (t_3, m_3)\big)\\
&\quad= (t_1, m_1)\,{}_a\bullet_b(t_2\,{}_c\bullet_d\,t_3,\, t_2\,{}_c\overset{L}{\bullet}_d\,m_3 + m_2\,{}_c\overset{R}{\bullet}_d\,t_3)\\
&\quad= \big(t_1\,{}_a\bullet_b(t_2\,{}_c\bullet_d\,t_3),\, t_1\,{}_a\overset{L}{\bullet}_b(t_2\,{}_c\overset{L}{\bullet}_d\,m_3) + t_1\,{}_a\overset{L}{\bullet}_b(m_2\,{}_c\overset{R}{\bullet}_d\,t_3) + m_1\,{}_a\overset{R}{\bullet}_b(t_2\,{}_c\bullet_d\,t_3)\big),
\end{aligned}$$

while

$$\begin{aligned}
&-{}_c\bullet_d(\,{}_a\bullet_b \otimes \mathbb{1})\big((t_1, m_1) \otimes (t_2, m_2) \otimes (t_3, m_3)\big)\\
&\quad= -(t_1\,{}_a\bullet_b\,t_2,\, t_1\,{}_a\overset{L}{\bullet}_b\,m_2 + m_1\,{}_a\overset{R}{\bullet}_b\,t_2)\,{}_c\overset{L}{\bullet}_d(t_3, m_3)\\
&\quad= -\big((t_1\,{}_a\bullet_b\,t_2)\,{}_c\bullet_d\,t_3,\, (t_1\,{}_a\bullet_b\,t_2)\,{}_c\overset{L}{\bullet}_d\,m_3 + (t_1\,{}_a\overset{L}{\bullet}_b\,m_2)\,{}_c\overset{R}{\bullet}_d\,t_3 + (m_1\,{}_a\overset{R}{\bullet}_b\,t_2)\,{}_c\overset{R}{\bullet}_d\,t_3\big).
\end{aligned}$$

On the other hand, one has

$$t_1\, {}_a\bullet_b (t_2\, {}_c\bullet_d\, t_3) = -(t_1\, {}_a\bullet_b\, t_2)\, {}_c\bullet_d\, t_3$$

by (6.80),

$$t_1\, {}_a\overset{L}{\bullet}_b (t_2\, {}_c\overset{L}{\bullet}_d\, m_3) = -(t_1\, {}_a\bullet_b\, t_2)\, {}_c\overset{L}{\bullet}_d\, m_3$$

by (7.1),

$$t_1\, {}_a\overset{L}{\bullet}_b (m_2\, {}_c\overset{R}{\bullet}_d\, t_3) = -(t_1\, {}_a\overset{L}{\bullet}_b\, m_2)\, {}_c\overset{R}{\bullet}_d\, t_3$$

by (7.3), and

$$m_1\, {}_a\overset{R}{\bullet}_b (t_2\, {}_c\bullet_d\, t_3) = -(m_1\, {}_a\overset{R}{\bullet}_b\, t_2)\, {}_c\overset{R}{\bullet}_d\, t_3$$

by (7.4). The requisite equality (6.80) follows immediately.

Definition 7.4 Let $\mathscr{T}$ be an odd modular operad and M a $\mathscr{T}$-module with the structure operations ${}_a\overset{L}{\bullet}_b$ and $\blacklozenge_{ab}$. We define the modular module sM by

$$\mathsf{sM}(S; g) := \uparrow \mathsf{M}(S; g),\ (S; g) \in \mathtt{Cor} \times \mathbb{A},$$

with the actions (6.48) given as $\mathsf{sM}(\sigma) := \uparrow \mathsf{M}(\sigma) \downarrow$. For disjoint finite sets S_1, S_2, symbols a, b and genera $g_1, g_2 \in \mathbb{A}$ we define degree $+1$ morphisms

$${}_a\overset{L}{\underline{\bullet}}_b : \mathscr{T}\big(S_1 \sqcup \{a\}; g_1\big) \otimes \mathsf{sM}\big(S_2 \sqcup \{b\}; g_2\big) \to \mathsf{sM}(S_1 \sqcup S_2; g_1 + g_2)$$

by

$${}_a\overset{L}{\underline{\bullet}}_b := -\uparrow {}_a\overset{L}{\bullet}_b(\mathbb{1} \otimes \downarrow). \tag{7.7}$$

For a finite set S, genus $g \in \mathbb{A}$, and symbols u, v we also define degree $+1$ maps

$$\underline{\blacklozenge}_{uv} = \underline{\blacklozenge}_{vu} : \mathsf{sM}\big(S \sqcup \{u, v\}; g\big) \to \mathsf{sM}(S; g + s)$$

by the formula

$$\underline{\blacklozenge}_{uv} := -\uparrow \blacklozenge_{uv} \downarrow. \tag{7.8}$$

The auxiliary "right" actions ${}_a\overset{R}{\underline{\bullet}}_b : \mathsf{sM} \otimes \mathscr{T} \to \mathsf{sM}$ of the suspension sM are given by ${}_a\overset{R}{\underline{\bullet}}_b := -\uparrow {}_a\overset{R}{\bullet}_b(\downarrow \otimes \mathbb{1})$.

The $\mathscr{T}$-module sM is called the *suspension* of M. The *desuspension* $\mathsf{s}^{-1}\mathsf{M}$ is defined analogously. We define $\mathsf{s}^k\mathsf{M}$ for an arbitrary integer $k \in \mathbb{Z}$ by iteration.

One has the expected:

Lemma 7.3 *The modular module* sM *with the above operations is a* $\mathscr{T}$*-module.*

Proof. A straightforward verification. Let us, for instance, check axiom (vii) of Definition 7.1. We have, by the definition of the structure operations of sM and by (7.2) for the suspension M,

$$\begin{aligned} {}_a\overset{L}{\underline{\bullet}}_b(\bullet_{uv} \otimes \mathbb{1}) &= -\downarrow {}_a\overset{L}{\bullet}_b(\mathbb{1} \otimes \uparrow)(\bullet_{uv} \otimes \mathbb{1}) = \downarrow {}_a\overset{L}{\bullet}_b(\bullet_{uv} \otimes \mathbb{1})(\mathbb{1} \otimes \uparrow) \\ &= -\downarrow \blacklozenge_{uv}\ {}_a\overset{L}{\bullet}_b(\mathbb{1} \otimes \uparrow). \end{aligned}$$

Similarly,

$$-\underline{\blacklozenge}_{uv}\ {}_a\overset{L}{\underline{\bullet}}_b = -\downarrow \blacklozenge_{uv} \uparrow \downarrow {}_a\overset{L}{\bullet}_b(\mathbb{1} \otimes \uparrow) = -\downarrow \blacklozenge_{uv}\ {}_a\overset{L}{\bullet}_b(\mathbb{1} \otimes \uparrow),$$

so ${}_a\overset{L}{\underline{\bullet}}_b(\bullet_{uv} \otimes \mathbb{1}) = -\underline{\blacklozenge}_{uv}\ {}_a\overset{L}{\underline{\bullet}}_b$ as required.

Remark 7.2 The minus sign in (7.7) resp. (7.8) defining ${}_a\overset{L}{\underline{\bullet}}_b$ resp. $\underline{\blacklozenge}_{ab}$ is crucial for establishing axioms (iii) and (vii) of Definition 7.1 for sM. These axioms are not homogeneous with respect to the number of operations, so they are not invariant under the change

$${}_a\overset{L}{\underline{\bullet}}_b \mapsto -\ {}_a\overset{L}{\underline{\bullet}}_b \text{ and/or } \underline{\blacklozenge}_{uv} \mapsto -\underline{\blacklozenge}_{uv}.$$

The other axioms are not sign-sensitive. We encourage the reader to analyze the proof of Lemma 7.3 and verify that without the minus sign in (7.8), the equality ${}_a\overset{L}{\underline{\bullet}}_b(\bullet_{uv} \otimes \mathbb{1}) = -\underline{\blacklozenge}_{uv}\ {}_a\overset{L}{\underline{\bullet}}_b$ would not hold.

Lemmas 7.2 and 7.3 imply that $\mathscr{T} \oplus \mathsf{s}^k\mathsf{M}$ is an odd modular operad for each integer k. The following lemma characterizes derivations via homomorphisms.

Lemma 7.4 *Let* $\theta : \mathscr{T} \to \mathsf{M}$ *a degree k morphism of modular modules. Associate to it a degree* 0 *morphism* $\overline{\theta} : \mathscr{T} \to \mathsf{s}^{-k}\mathsf{M}$ *of modular modules by*

$$\overline{\theta} := \downarrow^k \theta$$

and another degree 0 *morphism* $\Theta : \mathscr{T} \to \mathscr{T} \oplus \mathsf{s}^{-k}\mathsf{M}$ *of modular modules given by*

$$\Theta := (\mathbb{1}_{\mathscr{T}}, \overline{\theta}).$$

Then the following three statements are equivalent:

(i) θ *is a degree k derivation,*
(ii) $\overline{\theta}$ *is a degree 0 derivation, and*
(iii) Θ *is a morphism of odd modular operads.*

Proof. Applying the iterated desuspension $\downarrow^k$ to both sides of the Leibniz rule (7.5), we obtain

$$\downarrow^k\theta\ {}_a\bullet_b = (-1)^k\big(\downarrow^k {}_a\overset{R}{\bullet}_b(\uparrow^k \otimes \mathbb{1})(\downarrow^k\theta \otimes \mathbb{1}) + \downarrow^k {}_a\overset{L}{\bullet}_b(\mathbb{1} \otimes \uparrow^k)(\mathbb{1} \otimes \downarrow^k\theta)\big). \quad (7.9)$$

Taking into account that the structure operations ${}_a\overset{L}{\bullet}_b$ of M and the structure operations ${}_a\overset{L}{\underline{\bullet}}_b$ of the iterated desuspension $\mathsf{s}^{-k}\mathsf{M}$ are related by the iteration of (7.7) as

$${}_a\overset{L}{\underline{\bullet}}_b = (-1)^k \downarrow^k {}_a\overset{L}{\bullet}_b(\mathbb{1}\otimes \uparrow^k)$$

and that, likewise,

$${}_a\overset{R}{\underline{\bullet}}_b = (-1)^k \downarrow^k {}_a\overset{R}{\bullet}_b(\uparrow^k \otimes \mathbb{1})$$

we see that (7.9) is satisfied if and only if $\overline{\theta}$ fulfills

$$\overline{\theta}\ {}_a\bullet_b = {}_a\overset{R}{\underline{\bullet}}_b(\overline{\theta} \otimes \mathbb{1}) + {}_a\overset{L}{\underline{\bullet}}_b(\mathbb{1} \otimes \overline{\theta}). \quad (7.10)$$

Similarly, applying $\downarrow^k$ to both sides of (7.6) gives

$$\downarrow^k\theta\bullet_{uv} = (-1)^k\downarrow^k\blacklozenge_{uv}\uparrow^k\downarrow^k\theta.$$

which is equivalent to

$$\overline{\theta}\bullet_{uv} = \underline{\blacklozenge}_{uv}\overline{\theta}. \quad (7.11)$$

Since (7.10) together with (7.11) means that $\overline{\theta}$ is a degree 0 derivation, we proved the equivalence of (i) and (ii).

Let us check that (ii) is satisfied if and only if $\Theta = (\mathbb{1}, \overline{\theta})$ is an operad morphism, i.e. that

$$\big(t_1\ {}_a\bullet_b\ t_2, \overline{\theta}(t_1\ {}_a\bullet_b\ t_2)\big) = \big(t_1, \overline{\theta}(t_1)\big)\ {}_a\bullet_b\ \big(t_2, \overline{\theta}(t_2)\big) \quad (7.12)$$

for $t_1 \in \otimes\big(S_1 \sqcup \{a\}; g_1\big)$, $t_2 \in \otimes\big(S_2 \sqcup \{a\}; g_2\big)$, and

$$\big(\bullet_{uv}\, t, \overline{\theta} \bullet_{uv} (t)\big) = \bullet_{uv}\big(t, \overline{\theta}(t)\big) \tag{7.13}$$

for $t \in \mathscr{T}\big(S \sqcup \{u, v\}; g\big)$.

By the definition of the operad structure of $\mathscr{T} \oplus \mathsf{s}^{-k}\mathsf{M}$, the right-hand side of (7.12) equals

$$\big(t_1 \,{}_a\bullet_b\, t_2, t_1 \,{}_a\overset{L}{\underline{\bullet}}_b\, \overline{\theta}(t_2) + \overline{\theta}(t_1) \,{}_a\overset{R}{\underline{\bullet}}_b\, t_2\big).$$

Since the degree of $\overline{\theta}$ is 0, the second component of the above expression is precisely the right-hand side of (7.10) evaluated at $t_1 \otimes t_2$, so (7.12) holds. Similarly, the right-hand side of (7.13) equals $((\bullet_{uv} t, \underline{\blacklozenge}_{uv} \overline{\theta}(t)))$, thus (7.11) is apparently equivalent to (7.13). This shows that (ii) is equivalent to (iii) and finishes the proof of the lemma.

Theorem 7.1 below characterizes derivations of the free odd modular operad $\widetilde{\mathbb{F}}(E)$ generated by a modular module E. Let us recall that the freeness of $\widetilde{\mathbb{F}}(E)$ means that, for each odd modular operad $\mathscr{T}$ and a morphism $f : E \to \mathscr{T}$ of modular modules, there exists a unique morphism Φ of odd modular operads such that the diagram

$$\begin{array}{ccc} E & \xrightarrow{\ f\ } & \mathscr{T} \\ {\scriptstyle\iota}\downarrow & \nearrow_{\Phi} & \\ \widetilde{\mathbb{F}}(E) & & \end{array}$$

in which $\iota : E \to \widetilde{\mathbb{F}}(E)$ is the inclusion, commutes.

Theorem 7.1 *Let* M *be a module over the free odd modular operad* $\widetilde{\mathbb{F}}(E)$*. The restriction* $\theta \longmapsto \theta|_E$ *to the space of generators defines a one-to-one correspondence between degree* k *derivations* $\theta : \widetilde{\mathbb{F}}(E) \to \mathsf{M}$ *and degree* k *modular module morphisms* $\vartheta : E \to \mathsf{M}$*.*

Proof. To simplify the arguments, we start by showing that we may assume without loss of generality that $k = 0$. By Lemma 7.4 with $\mathscr{T} = \widetilde{\mathbb{F}}(E)$, there is a one-to-one correspondence between degree k derivations $\theta : \widetilde{\mathbb{F}}(E) \to \mathsf{M}$ and degree 0 derivations $\overline{\theta} : \widetilde{\mathbb{F}}(E) \to \mathsf{s}^{-k}\mathsf{M}$, given by $\overline{\theta} :=\uparrow^{-k}\theta$. Likewise, there is a one-to-one correspondence between degree k modular module morphisms $\vartheta : E \to \mathsf{M}$ and degree 0 modular module morphisms $\overline{\vartheta} : E \to \mathsf{s}^{-k}\mathsf{M}$, given by $\overline{\vartheta} :=\uparrow^{-k}\vartheta$. It is clear that, if $\vartheta = \theta|_E$, then $\overline{\vartheta} = \overline{\theta}|_E$. We may therefore replace, in Theorem 7.1, θ by $\overline{\theta}$, ϑ by $\overline{\vartheta}$ and M by $\mathsf{s}^{-k}\mathsf{M}$ and thus assume that $k = 0$.

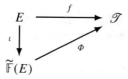

By Lemma 7.4 again, there is a one-to-one correspondence between degree 0 derivations $\theta : \widetilde{\mathbb{F}}(E) \to \mathsf{M}$ and morphism of modular operads

$$\Theta := \big(\mathbb{1}_{\widetilde{\mathbb{F}}(E)}, \theta\big) : \widetilde{\mathbb{F}}(E) \to \widetilde{\mathbb{F}}(E) \oplus \mathsf{M}. \tag{7.14}$$

By the freeness of $\widetilde{\mathbb{F}}(E)$, such a morphism is determined by its restriction

$$\Theta|_E = \big(\iota, \theta|_E\big) : E \to \widetilde{\mathbb{F}}(E) \oplus \mathsf{M}$$

which is in turn determined by the restriction $\theta|_E : E \to \mathsf{M}$. It is therefore enough to show that every modular module morphism $\vartheta : E \to \mathsf{M}$ extends to a modular operad morphism of as in (7.14), determining a derivation that restricts to it.

Let $\Theta : \widetilde{\mathbb{F}}(E) \to \widetilde{\mathbb{F}}(E) \oplus \mathsf{M}$ be the unique extension of the modular module morphism $(\iota, \vartheta) : E \to \widetilde{\mathbb{F}}(E) \oplus \mathsf{M}$. This extension is necessary of the form $\Theta = (\Phi, \theta)$ for some $\Phi : \widetilde{\mathbb{F}}(E) \to \widetilde{\mathbb{F}}(E)$. We must show that $\Phi = \mathbb{1}_{\widetilde{\mathbb{F}}(E)}$.

The projection $\mathrm{pr}_1 : \widetilde{\mathbb{F}}(E) \oplus \mathsf{M} \to \widetilde{\mathbb{F}}(E)$ to the first summand is clearly an operad morphism, and so is the composition $\Phi = \mathrm{pr}_1 \circ \Theta$. By construction, $\Phi|_E = \iota$. Since the identity $\mathbb{1}_{\widetilde{\mathbb{F}}(E)}$ also restricts to ι, Φ must equal $\mathbb{1}_{\widetilde{\mathbb{F}}(E)}$ by the uniqueness of the extension. The equality $\theta|_E = \vartheta$ is obvious.

7.2 Feynman Transform

In this section we define the cobar construction for modular (co)operads, which is called in this context the Feynman transform. To understand the idea of this construction better, we recall very briefly the cobar construction for the classical non-counital coassociative coalgebras.

If C is such a coalgebra with the comultiplication $\Delta : C \to C \otimes C$, then its *cobar construction* $\Omega(C) = (\mathbb{T}(\uparrow C), \partial)$ is the tensor algebra $\mathbb{T}(\uparrow C)$ on the suspension of C, with the differential that is the unique extension of the linear degree $+1$ map

$$(\uparrow \otimes \uparrow) \circ \Delta \circ \downarrow : \uparrow C \to \uparrow C \otimes \uparrow C$$

into a degree $+1$ derivation. Such an unique extension exists since $\mathbb{T}(\uparrow C)$ is the free associative algebra generated by $\uparrow C$. A straightforward calculation shows that the coassociativity of Δ implies $\partial^2 = 0$. The cobar construction is therefore a dg-associative algebra.

If C is a dg-coalgebra with a differential d_C, then $\uparrow d_C \downarrow : \uparrow C \to \uparrow C$ extends into a degree $+1$ derivation $d : \mathbb{T}(\uparrow C) \to \mathbb{T}(\uparrow C)$. One easily proves that

$$\partial^2 = d^2 = d\partial + \partial d = 0. \tag{7.15}$$

There is a modification of the above construction that avoids the use of the suspension. In Example 6.30 we recalled anti-associative algebras; let us denote by $\widetilde{\mathbb{T}}(C)$ the free anti-associative algebra generated by a graded vector space C and $\iota : C \hookrightarrow \widetilde{\mathbb{T}}(C)$ the canonical inclusion. The algebra $\widetilde{\mathbb{T}}(C)$ can be realized as the tensor algebra $\mathbb{T}(C)$ with suitably redefined degrees and the tensor product taken with an appropriate sign; we leave the details to the interested reader.

If $C = (C, \Delta)$ is a coalgebra as above, we define its "odd" cobar construction $\widetilde{\Omega}(C)$ as the couple $(\widetilde{\mathbb{T}}(C), \partial)$, where ∂ is the unique degree $+1$ derivation extending the degree $+1$ map

$$C \xrightarrow{\Delta} C \otimes C \xrightarrow{\iota\otimes\iota} \widetilde{\mathbb{T}}(C) \otimes \widetilde{\mathbb{T}}(C) \xrightarrow{\star} \widetilde{\mathbb{T}}(C).$$

We leave as an exercise to prove that $\partial^2 = 0$, so $\widetilde{\Omega}(C)$ is a dg-anti-associative algebra. As before, a differential d_C of C induces a second differential $d : \widetilde{\Omega}(C) \to \widetilde{\Omega}(C)$ such that (7.15) is satisfied.

Notice that if A is an dg-associative algebra, then its desuspension $\downarrow A$ has a natural structure of an anti-associative algebra. Likewise, suspensions of dg-anti-associative algebras are ordinary associative algebras. This correspondence defines an equivalence between the category of dg-associative algebras and the category of dg-anti-associative algebras. Since the ordinary and the odd cobar constructions correspond to each other under this correspondence, both constructions are essentially equivalent.

As the classical cobar construction recalled above acts on coassociative coalgebras, the Feynman transform acts on modular *co*operads. The underlying objects of modular cooperads are modular comodules:

Definition 7.5 A *modular comodule* is a contravariant functor

$$E : \mathtt{Cor} \times \mathbb{A} \to \mathtt{Chain}.$$

Definition 7.6 Each modular comodule E has its *associated modular module*

$$\mathring{E} : \mathtt{Cor} \times \mathbb{A} \to \mathtt{Chain}$$

with $\mathring{E}(S; g) := E(S; g)$ and $\mathring{E}(\sigma) : \mathring{E}(S; g) \to \mathring{E}(T; g)$ defined, for an isomorphism $\sigma : S \xrightarrow{\cong} T$, by $\mathring{E}(\sigma) := E(\sigma^{-1})$.

The assignment $E \mapsto \mathring{E}$ defines an isomorphism between the category of modular comodules and the category of modular modules.

Definition 7.7 A *modular cooperad* with step s consists of a modular comodule

$$\mathscr{C} = \{\mathscr{C}(S; g) \in \mathtt{Chain};\ (S; g) \in \mathtt{Cor} \times \mathbb{A}\}$$

together with degree 0 morphisms (cocompositions)

$$ {}_{S_1}{\circ}^{a\,b}_{S_2} = {}_{S_1;\,g_1}{\circ}^{a\,b}_{S_2;\,g_2} : \mathscr{C}(S_1 \sqcup S_2;\, g) \to \mathscr{C}(S_1 \sqcup \{a\};\, g_1) \otimes \mathscr{C}(S_2 \sqcup \{b\};\, g_2) \tag{7.16} $$

defined for arbitrary disjoint finite sets S_1, S_2, symbols a, b, and genera $g_1, g_2 \in \mathbb{A}$ such that $g_1 + g_2 = g$. There are, moreover, degree 0 morphisms (cocontractions)

$$ \circ^{uv}_g = \circ^{vu}_g = \circ^{uv} : \mathscr{C}(S;\, g + s) \to \mathscr{C}(S \sqcup \{u, v\};\, g) \tag{7.17} $$

given for any finite set S, genus $g \in \mathbb{A}$, and symbols u, v. These data must satisfy the following axioms:

(i) For arbitrary isomorphisms $\rho : S_1 \sqcup \{a\} \to T_1$ and $\sigma : S_2 \sqcup \{b\} \to T_2$ of finite sets and genera $g_1, g_2 \in \mathbb{A}$, one has the equality

$$ \big(\mathscr{C}(\rho) \otimes \mathscr{C}(\sigma)\big)\, {}_{T_1 \setminus \rho(a)}{\circ}^{\rho(a)\,\rho(b)}_{T_2 \setminus \sigma(a)} = {}_{S_1}{\circ}^{a\,b}_{S_2}\, \mathscr{C}(\rho|_{S_1} \sqcup \sigma|_{S_2}) $$

of maps

$$ \mathscr{C}\big(T_1 \sqcup T_2 \setminus \{\rho(a), \sigma(b)\};\, g_1 + g_2\big) \to \mathscr{C}\big(S_1 \sqcup \{a\};\, g_1\big) \otimes \mathscr{C}\big(S_2 \sqcup \{b\};\, g_2\big). $$

(ii) For each isomorphism $\rho : S \sqcup \{u, v\} \to T$ of finite sets and a genus $g \in \mathbb{A}$, one has the equality

$$ \mathscr{C}(\rho)\, \circ^{\rho(a)\rho(b)} = \circ^{ab}\, \mathscr{C}(\rho|_S) \tag{7.18} $$

of maps $\mathscr{C}\big(T \setminus \{\rho(u), \rho(v)\};\, g + s\big) \to \mathscr{C}\big(S \sqcup \{u, v\};\, g\big)$.

(iii) For S_1, S_2, a, b and g_1, g_2 as in (7.16), one has the equality

$$ \tau\, {}_{b}{\circ}^{S_2\,S_1}_{a} = {}_{S_1}{\circ}^{a\,b}_{S_2} \tag{7.19} $$

of maps $\mathscr{C}\big(S_1 \sqcup S_2;\, g_1 + g_2\big) \to \mathscr{C}(S_1 \sqcup \{a\};\, g_1) \otimes \mathscr{C}(S_2 \sqcup \{b\};\, g_2)$.

(iv) For mutually disjoint sets S_1, S_2, S_3, symbols a, b, c, d and genera $g_1, g_2, g_3 \in \mathbb{A}$, one has the equality

$$ (\mathbb{1} \otimes {}_{S_2 \sqcup \{b\}}{\circ}^{c\,d}_{S_3})\, {}_{S_1}{\circ}^{a\,b}_{S_2 \sqcup S_3} = ({}_{S_1}{\circ}^{a\,b}_{S_2 \sqcup \{c\}} \otimes \mathbb{1})\, {}_{S_1 \sqcup S_2}{\circ}^{c\,d}_{S_3} \tag{7.20} $$

of maps from $\mathscr{C}\big(S_1 \sqcup S_2 \sqcup S_3;\, g_1 + g_2 + g_3\big)$ to the space

$$ \mathscr{C}\big(S_1 \sqcup \{a\};\, g_1\big) \otimes \mathscr{C}\big(S_2 \sqcup \{b, c\};\, g_2\big) \otimes \mathscr{C}\big(S_3 \sqcup \{d\};\, g_3\big). $$

(v) For a finite set S, symbols a, b, c, d and a genus $g \in \mathbb{A}$ one has the equality

$$\circ^{ab} \circ^{cd} = \circ^{cd} \circ^{ab} \tag{7.21}$$

of maps $\mathscr{C}(S; g + 2s) \to \mathscr{C}\big(S \sqcup \{a, b, c, d\}; g\big)$.

(vi) For finite sets S_1, S_2, symbols a, b, c, d and genera $g_1, g_2 \in \mathbb{A}$, one has the equality

$$ {}_{S_1 \sqcup \{c\}}{}^{a}\circ^{b}{}_{S_2 \sqcup \{d\}} \circ^{cd} = {}_{S_1 \sqcup \{a\}}{}^{c}\circ^{d}{}_{S_2 \sqcup \{b\}} \circ^{ab} \tag{7.22}$$

of maps $\mathscr{C}(S_1 \sqcup S_2; g_1 + g_2 + s) \to \mathscr{C}\big(S_1 \sqcup \{a, c\}; g_1\big) \otimes \mathscr{C}\big(S_2 \sqcup \{b, d\}; g_2\big)$.

(vii) For finite sets S_1, S_2, symbols a, b, u, v, and genera $g_1, g_2 \in \mathbb{A}$, one has the equality

$$(\circ^{uv} \otimes \mathbb{1})\ {}_{S_1}{}^{a}\circ^{b}{}_{S_2} = {}_{S_1 \sqcup \{u, v\}}{}^{a}\circ^{b}{}_{S_2} \circ^{uv} \tag{7.23}$$

of maps $\mathscr{C}(S_1 \sqcup S_2; g_1 + g_2 + s) \to \mathscr{C}\big(S_1 \sqcup \{a, u, v\}; g_1\big) \otimes \mathscr{C}\big(S_2 \sqcup \{b\}; g_2\big)$.

Convention From this moment on we will assume that the semigroup $\mathbb{A}$ is such that the set

$$\{(g_1, g_2) \in \mathbb{A}^{\times 2} \mid g = g_1 + g_2\} \tag{7.24}$$

is finite for each $g \in \mathbb{A}$.

The finiteness (7.24) guarantees that some constructions or formulas work without the necessity to pass to completions. Denote, for instance, for a finite set S,

$$\mathscr{C}(S) := \bigoplus_{g \in \mathbb{A}} \mathscr{C}(S; g).$$

It is a bigraded vector space, with the first grading given by the grading of the graded vector spaces $\mathscr{C}(S; g)$, and the second grading given by the genus. Thanks to the finiteness of the sets (7.24), the maps (7.16) assemble into a bidegree-$(0, 0)$ map $\mathscr{C}(S_1 \sqcup S_2) \to \mathscr{C}(S_1 \sqcup \{a\}) \otimes \mathscr{C}(S_2 \sqcup \{b\})$. Notice that the finiteness is always fulfilled when $\mathbb{A} = \mathbb{N}$, which is the case of the most important applications.

Condition (7.24) could be replaced by a weaker one. We may, e.g. assume that for each S_1, S_2, and g as in (7.16) is the set

$$\Big\{(g_1, g_2) \in \mathbb{A}^{\times 2},\ g_1 + g_2 = g \ \Big|\ {}_{S_1; g_1}{}^{a}\circ^{b}{}_{S_2; g_2} \neq 0\Big\}$$

finite, and make a similar assumption also about the cocontractions $\circ_g^{uv}$. In all applications we know (7.24) is however satisfied.

The main source of examples of modular cooperads are the piecewise linear duals of modular operads satisfying suitable finiteness conditions.

Definition 7.8 A $\mathbb{Z}$-graded vector space V is of *finite type* if each component V^k, $k \in \mathbb{Z}$, is finite-dimensional. It is *finite-dimensional* if the associated total space

$$V = \bigoplus_{k \in \mathbb{Z}} V^k$$

is finite-dimensional. A modular operad $\mathcal{M}$ is of *finite type* (resp. *finite-dimensional*) if the graded vector space $\mathcal{M}(S; g)$ is such for each $(S, g) \in \mathtt{Cor} \times \mathbb{A}$.

Proposition 7.1 *Let $\mathcal{M}$ be a modular operad of finite type. For $k_1, k_2 \in \mathbb{Z}$, $g_1, g_2 \in \mathbb{A}$ and $S_1, S_2 \in \mathtt{Cor}$ denote by*

$${}^{g_1,k_1}{}_a\circ_b^{g_2,k_2} : \mathcal{M}\big(S_1 \sqcup \{a\}; g_1\big)^{k_1} \otimes \mathcal{M}\big(S_2 \sqcup \{b\}; g_2\big)^{k_2} \to \mathcal{M}(S_1 \sqcup S_2; g_1 + g_2)^{k_1+k_2}$$

the restriction of the structure operation (6.51) to the indicated components. Assume that the set

$$\big\{(k_1, k_2) \in \mathbb{Z}^{\times 2},\ k_1 + k_2 = k \mid {}^{k_1,g_1}{}_a\circ_b^{k_2,g_2} \neq 0\big\} \tag{7.25}$$

is finite for each $S_1, S_2, g_1, g_2 \in \mathbb{A}$ and $k \in \mathbb{Z}$. Then the modular module

$$\mathcal{M}^{\#} = \big\{\mathcal{M}(S; g)^{\#} \in \mathtt{Chain} \,\big|\, (S, g) \in \mathtt{Cor} \times \mathbb{A}\big\}$$

of piecewise linear duals has a natural modular cooperad structure induced from the modular operad structure of $\mathcal{M}$.

Remark 7.3 The finiteness of (7.25) is always satisfied when $\mathcal{M}$ is non-negatively or non-positively graded, or finite-dimensional,

Proof (of Proposition 7.1) To shorten the notation, we denote for $k \in \mathbb{Z}$, $g \in \mathbb{A}$ and $S \in \mathtt{Cor}$ by $\mathcal{C}(S; g)^k$ the linear dual $\big(\mathcal{M}(S; g)^k\big)^{\#}$ of the degree-k graded component of $\mathcal{M}(S; g)$. It is clear that the collection

$$\mathcal{M}^{\#} = \mathcal{C} = \big\{\mathcal{C}(S; g) \in \mathtt{Chain} \,\big|\, (S, g) \in \mathtt{Cor} \times \mathbb{A}\big\}$$

with

$$\mathcal{C}(S; g) = \mathcal{M}(S; g)^{\#} = \bigoplus_{k \in \mathbb{Z}} \mathcal{C}(S; g)^k$$

and the induced differential $d^{\#}_{\mathscr{M}}$ is a modular comodule. Consider the linear dual

$$({}^{g_1,k_1}{}_a\circ_b^{g_2,k_2})^{\#} : \mathscr{C}(S_1 \sqcup S_2; g)^k \longrightarrow \big(\mathscr{M}\big(S_1 \sqcup \{a\}; g_1\big)^{k_1} \otimes \mathscr{M}\big(S_2 \sqcup \{b\}; g_2\big)^{k_2}\big)^{\#}$$

of the restrictions ${}^{g_1,k_1}{}_a\circ_b^{g_2,k_2}$. Since $\mathscr{M}$ is of finite type by assumption, the canonical inclusion

$$\begin{aligned} {}^{g_1,k_1}\iota^{g_2,k_2} : \mathscr{C}(S_1 \sqcup \{a\}; g_1)^{k_1} \otimes \mathscr{C}(S_1 \sqcup \{b\}; g_2)^{k_2} \hookrightarrow \\ \hookrightarrow \big(\mathscr{M}\big(S_1 \sqcup \{a\}; g_1\big)^{k_1} \otimes \mathscr{M}\big(S_2 \sqcup \{b\}; g_2\big)^{k_2}\big)^{\#} \end{aligned}$$

is an isomorphism, so one can define a map

$$({}_{S_1;\,g_1}{}^a\circ^b{}_{S_2;g_2})^k : \mathscr{C}(S_1 \sqcup S_2; g)^k \to \prod_{k_1+k_2=k} \mathscr{C}\big(S_1 \sqcup \{a\}; g_1\big)^{k_1} \otimes \mathscr{C}\big(S_2 \sqcup \{b\}; g_2\big)^{k_2}$$

as the product

$$({}_{S_1;\,g_1}{}^a\circ^b{}_{S_2;g_2})^k := \prod_{k_1+k_2=k} ({}^{g_1,k_1}\iota^{g_2,k_2})^{-1}({}^{g_1,k_1}{}_a\circ_b^{g_2,k_2})^{\#}. \tag{7.26}$$

By the finiteness of (7.25), the product in (7.26) has only finitely many nontrivial components, so $({}_{S_1;\,g_1}{}^a\circ^b{}_{S_2;g_2})^k$ is in fact a map

$$({}_{S_1;\,g_1}{}^a\circ^b{}_{S_2;g_2})^k : \mathscr{C}(S_1 \sqcup S_2; g)^k \to \bigoplus_{k_1+k_2=k} \mathscr{C}\big(S_1 \sqcup \{a\}; g_1\big)^{k_1} \otimes \mathscr{C}\big(S_2 \sqcup \{b\}; g_2\big)^{k_2}$$

which is the kth component of a degree-0 map

$${}_{S_1;\,g_1}{}^a\circ^b{}_{S_2;g_2} : \mathscr{C}(S_1 \sqcup S_2; g) \to \mathscr{C}\big(S_1 \sqcup \{a\}; g_1\big) \otimes \mathscr{C}\big(S_2 \sqcup \{b\}; g_2\big). \tag{7.27}$$

The operations $\circ_g^{uv} =: \mathscr{C}(S; g+s) \to \mathscr{C}(S \sqcup \{u, v\}; g)$ are just simple-minded duals of the contractions (6.52), the dualization here presents no problem. Since the axioms of modular cooperads are the exact formal duals of the axioms of modular operads, the modular comodule $\mathscr{M}^{\#} = \mathscr{C}$ with the structure operations (7.27) and $\circ_g^{uv} := {\circ_{uv}}^{\#}$ form a modular cooperad.

Let us try to define the Feynman transform of a modular cooperad by mimicking the definition of the cobar construction of a coassociative coalgebra recalled at the beginning of this section. One is tempted to take the free modular operad $\mathbb{F}(\uparrow \mathscr{C})$ on the component-wise suspension of a modular cooperad $\mathscr{C}$ and equip it with a differential that extends the structure operations of $\mathscr{C}$ into a degree $+1$ derivation of $\mathbb{F}(\uparrow \mathscr{C})$. Quite surprisingly, this would not work. The reason is that, while the cocompositions (7.16) define a degree $+1$ operations on this suspensions, the cocontractions (7.17) induce operations of degree 0! Fortunately, the analog of the alternative approach to the definition of the cobar construction using odd modular operads works.

In the following definition, $\widetilde{\mathbb{F}}(\mathring{\mathscr{C}})$ is the free odd modular operad generated by the modular module $\mathring{\mathscr{C}}$ associated with the underlying modular comodule of $\mathscr{C}$ in the correspondence of Definition 7.6, and $\iota : \mathring{\mathscr{C}} \to \widetilde{\mathbb{F}}(\mathring{\mathscr{C}})$ the canonical inclusion. Recall that each derivation $\theta : \widetilde{\mathbb{F}}(\mathring{\mathscr{C}}) \to \widetilde{\mathbb{F}}(\mathring{\mathscr{C}})$ is uniquely determined by its restriction $\theta\iota : \mathring{\mathscr{C}} \to \widetilde{\mathbb{F}}(\mathring{\mathscr{C}})$ along the natural inclusion $\iota : \mathring{\mathscr{C}} \to \widetilde{\mathbb{F}}(\mathring{\mathscr{C}})$ by Theorem 7.1.

Definition 7.9 Let $\mathscr{C}$ be a modular cooperad with the structure operations (7.16) and (7.17), and the internal differential $d_{\mathscr{C}}$. Let $\partial, d : \widetilde{\mathbb{F}}(\mathring{\mathscr{C}}) \to \widetilde{\mathbb{F}}(\mathring{\mathscr{C}})$ be degree $+1$ derivations defined, for $x \in \mathscr{C}(S; g) = \mathring{\mathscr{C}}(S; g)$, by the finite sum[2]

$$\partial\iota(x) := \sum_{g'+s=g} \bullet_{uv}\, \iota \circ^{uv}_{g'} (x) + \frac{1}{2} \sum_{A \sqcup B = S} \sum_{g_1+g_2=g} {}_a\bullet_b (\iota \otimes \iota)\, {}_{A;\, g_1}{}^{a}\circ^{b}{}_{B; g_2}(x) \tag{7.28}$$

where ${}_a\bullet_b$ and $\bullet_{uv}$ are the degree $+1$ structure operations of $\widetilde{\mathbb{F}}(\mathring{\mathscr{C}})$, and u, v, a, b are independent symbols, respectively, by

$$d\iota(x) := \iota d_{\mathscr{C}}(x).$$

The odd dg-modular operad $\mathscr{F}(\mathscr{C}) := \big(\widetilde{\mathbb{F}}(\mathring{\mathscr{C}}), \partial + d\big)$ is called *the Feynman transform* of the modular cooperad $\mathscr{C}$. If we need to distinguish between ∂ and d, we will call ∂ the *external* and d the *internal* differential.

Remark 7.4 Let us choose an element $s \in S$. It is clear that the rightmost term of (7.28) splits into the sum of two terms:

$$\sum_{\substack{A \sqcup B = S \\ s \in A}} \sum_{g_1+g_2=g} {}_a\bullet_b (\iota \otimes \iota)\, {}_{A;\, g_1}{}^{a}\circ^{b}{}_{B; g_2}(x) + \sum_{\substack{A \sqcup B = S \\ s \in B}} \sum_{g_1+g_2=g} {}_a\bullet_b (\iota \otimes \iota)\, {}_{A;\, g_1}{}^{a}\circ^{b}{}_{B; g_2}(x).$$

[2]In the first term in the right-hand side, the summation is not performed over the repeated indexes. All summations are finite by (7.24).

It moreover follows from the odd version of the symmetry (6.54) and from (7.19) that both terms of the above display are equal. Formula (7.28) can therefore be rewritten as

$$\partial\iota(x) := \sum_{g'+s=g} \bullet_{uv}\, \iota \circ^{uv}_{g'} (x) + \sum_{\substack{A\sqcup B=S\\ s\in A}} \sum_{g_1+g_2=g} {}_a\bullet_b(\iota\otimes\iota)\ {}_{A;\,g_1}{}^{a}\!\circ^{b}{}_{B;g_2}(x)$$

which does not use the rational one-half.

Theorem 7.2 *The derivations d and ∂ satisfy $\partial^2 = d^2 = \partial d + d\partial = 0$. In particular, $\partial + d$ is a differential, i.e. $(\partial + d)^2 = 0$.*

Proof (of Proposition 7.1) It follows from Lemma 7.1 that both ∂^2, d^2 and $\partial d + d\partial$ are degree $+2$ derivations. By Theorem 7.1, it suffices to verify the equalities $d^2 = 0$, $\partial d + d\partial = 0$ and $\partial^2 = 0$ on the generating space $\mathscr{C} = \mathrm{Im}(\iota) \subset \widetilde{\mathbb{F}}(\mathring{\mathscr{C}})$. For each finite set S, genus g and $x \in \mathscr{C}(S; g)$, one has by definition

$$d^2\iota(x) = d\iota(d_{\mathscr{C}}x) = \iota(d^2_{\mathscr{C}}x) = 0,$$

which proves that $d^2 = 0$. For the same x one has

$$d \bullet_{uv}\, \iota \circ^{uv}_{g'} (x) = - \bullet_{uv}\, d\iota \circ^{uv}_{g'} (x) = - \bullet_{uv}\, \iota d_{\mathscr{C}} \circ^{uv}_{g'} (x) = - \bullet_{uv}\, \iota \circ^{uv}_{g'} (d_{\mathscr{C}}x)$$

and

$$d\ {}_a\bullet_b(\iota\otimes\iota)\ {}_{A;\,g_1}{}^{a}\!\circ^{b}{}_{B;g_2}(x) = -\ {}_a\bullet_b(d\iota\otimes\iota + \iota\otimes d\iota)\ {}_{A;\,g_1}{}^{a}\!\circ^{b}{}_{B;g_2}(x)$$

$$= -\ {}_a\bullet_b(\iota\otimes\iota)\ {}_{A;\,g_1}{}^{a}\!\circ^{b}{}_{B;g_2}(d_{\mathscr{C}}x).$$

This shows that $d\partial\iota(x) = -\partial\iota(d_{\mathscr{C}}x)$ for each $x \in \mathscr{C}(S; g)$, so indeed $\partial d + d\partial = 0$.

Let us finally prove that $\partial^2 = 0$. To save the space, we will omit the summations over the genera whose presence will always be clear from the context. For $x \in \mathscr{C}(S; g)$ as above one has

$$\partial^2\iota(x) = \partial\Big(\bullet_{uv}\ \iota \circ^{uv} + \frac{1}{2} \sum_{A\sqcup B=S} {}_a\bullet_b(\iota\otimes\iota)\ {}_{A}^{a}\!\circ^{b}_{B}\Big)(x)$$

$$= - \bullet_{uv}\ \partial\iota \circ^{uv} (x) - \frac{1}{2} \sum_{A\sqcup B=S} {}_a\bullet_b(\partial\otimes\mathbb{1} + \mathbb{1}\otimes\partial)(\iota\otimes\iota)\ {}_{A}^{a}\!\circ^{b}_{B}(x)$$

$$= -\Big(\boxed{1} + \frac{1}{2}\boxed{2} + \frac{1}{2}\boxed{3} + \frac{1}{2}\boxed{4} + \frac{1}{2}\boxed{5} + \frac{1}{2}\boxed{6}\Big)(x) = 0,$$

where

$$\boxed{1} = \bullet_{ab}\,\bullet_{uv}\iota\,\circ^{uv}\,\circ^{ab},$$

$$\boxed{2} = \sum_{A\sqcup B=S\sqcup\{u,v\}} \bullet_{uv}\,{}_a\bullet_b(\iota\otimes\iota)\,{}^{a}_{A}\circ^{b}_{B}\,\circ^{uv},$$

$$\boxed{3} = \sum_{S_1\sqcup S_2=S} {}_a\bullet_b(\bullet_{uv}\otimes \mathbb{1})(\iota\otimes\iota)(\circ^{uv}\otimes\mathbb{1})\,{}^{a}_{S_1}\circ^{b}_{S_2},$$

$$\boxed{4} = \sum_{A\sqcup B=S}\ \sum_{A_1\sqcup A_2=A\sqcup\{a\}} {}_a\bullet_b({}_x\bullet_y\otimes\mathbb{1})(\iota\otimes\iota\otimes\iota)({}^{x}_{A_1}\circ^{y}_{A_2}\otimes\mathbb{1})\,{}^{a}_{A}\circ^{b}_{B},$$

$$\boxed{5} = \sum_{S_1\sqcup S_2=S} {}_a\bullet_b(\mathbb{1}\otimes\bullet_{uv})(\iota\otimes\iota)(\mathbb{1}\otimes\circ^{uv})\,{}^{a}_{S_1}\circ^{b}_{S_2}, \quad \text{and}$$

$$\boxed{6} = \sum_{A\sqcup B=S}\ \sum_{B_1\sqcup B_2=B\sqcup\{b\}} {}_a\bullet_b(\mathbb{1}\otimes{}_x\bullet_y)(\iota\otimes\iota\otimes\iota)(\mathbb{1}\otimes{}^{x}_{B_1}\circ^{y}_{B_2})\,{}^{a}_{A}\circ^{b}_{B}.$$

with formal variables a, b, u, v, x, y.

Let us start by analyzing $\boxed{1}$. Using the commutativity (7.21) in $\mathscr{C}$ and the anti-commutativity (6.81) in $\widetilde{\mathbb{F}}(\mathring{\mathscr{C}})$, we see that

$$\bullet_{ab}\,\bullet_{uv}\,\iota\,\circ^{uv}\,\circ^{ab} = -\,\bullet_{uv}\,\bullet_{ab}\iota\,\circ^{ab}\,\circ^{uv}.$$

Since a, b, u, v are formal independent variables, we may apply the substitution $a \leftrightarrow u$, $b \leftrightarrow v$ to the expression in the right-hand side and obtain

$$\bullet_{ab}\,\bullet_{uv}\,\iota\,\circ^{uv}\,\circ^{ab} = -\,\bullet_{ab}\,\bullet_{uv}\iota\,\circ^{uv}\,\circ^{ab}$$

which shows that $\boxed{1} = 0$.

More formally, define the automorphism $\rho : S\sqcup\{a,b,u,v\} \to S\sqcup\{a,b,u,v\}$ by

$$\rho|_S := \mathbb{1},\ \rho(a) := u \text{ and } \rho(b) := v.$$

Since $\mathring{\mathscr{C}}$ is a modular module, $\mathring{\mathscr{C}}(\rho)\mathring{\mathscr{C}}(\rho^{-1}) = \mathring{\mathscr{C}}(\mathbb{1}) = \mathbb{1}_{S\sqcup\{a,b,u,v\}}$, therefore

$$\iota = \widetilde{\mathbb{F}}(\mathring{\mathscr{C}})(\rho)\iota\mathring{\mathscr{C}}(\rho^{-1}) = \widetilde{\mathbb{F}}(\mathring{\mathscr{C}})(\rho)\iota\mathring{\mathscr{C}}(\rho),$$

one thus has

$$\bullet_{uv}\,\bullet_{ab}\,\iota\,\circ^{ab}\,\circ^{uv} = \bullet_{uv}\,\bullet_{ab}\,\widetilde{\mathbb{F}}(\mathring{\mathscr{C}})(\rho)\iota\mathring{\mathscr{C}}(\rho)\,\circ^{ab}\,\circ^{uv}.$$

The repeated use of (6.53) and (7.18) gives

$$\begin{aligned}
&\bullet_{uv} \bullet_{ab} \widetilde{\mathbb{F}}(\mathring{\mathscr{C}})(\rho)\iota\mathring{\mathscr{C}}(\rho) \circ^{ab} \circ^{uv} \\
&= \bullet_{uv} \widetilde{\mathbb{F}}(\mathring{\mathscr{C}})\big(\rho|_{S\sqcup\{a,b\}}\big) \bullet_{uv} \iota \circ^{uv} \mathring{\mathscr{C}}\big(\rho|_{S\sqcup\{a,b\}}\big)\circ^{uv} \\
&= \widetilde{\mathbb{F}}(\mathring{\mathscr{C}})\big(\rho|_{S}\big) \bullet_{ab} \bullet_{uv}\iota \circ^{uv} \circ^{ab}\mathring{\mathscr{C}}\big(\rho|_{S}\big) = \bullet_{ab} \bullet_{uv} \iota \circ^{uv} \circ^{ab},
\end{aligned}$$

so indeed

$$\bullet_{uv} \bullet_{ab} \iota \circ^{ab} \circ^{uv} = \bullet_{ab} \bullet_{uv} \iota \circ^{uv} \circ^{ab}.$$

Let us attend to $\boxed{2}$. There are four possibilities.

1. *Case* $u, v \in A$. Then $S = S_1 \sqcup S_2$, $A = S_1 \sqcup \{u, v\}$ and $B = S_2$ for some finite sets S_1, S_2, and $\boxed{2}$ equals

$$\sum_{S=S_1\sqcup S_2} \bullet_{uv}\, {}_a\bullet_b(\iota \otimes \iota)\ {}_{S_1 \sqcup \{u,v\}}{}^{a}\circ^{b}{}_{S_2} \circ^{uv}.$$

We rewrite this expression, using (6.83) and (7.23), as

$$-\sum_{S=S_1\sqcup S_2} {}_a\bullet_b\, (\circ_{uv} \otimes \mathbb{1})(\iota \otimes \iota)(\circ^{uv} \otimes \mathbb{1})\ {}_{S_1}{}^{a}\circ^{b}{}_{S_2},$$

which is $\boxed{3}$ with the minus sign.

2. *Case* $u, v \in B$. The mirror image of the previous one. By precisely the same arguments we obtain $\boxed{5}$ with the minus sign.
3. *Case* $u \in A$, $v \in B$. There clearly exist finite sets S_1, S_2 such that $A = S_1 \sqcup \{u\}$ and $B = S_2 \sqcup \{v\}$, so that $\boxed{2}$ can be rewritten as

$$\sum_{S=S_1\sqcup S_2} \bullet_{uv}\, {}_a\bullet_b(\iota \otimes \iota)\ {}_{S_1 \sqcup \{u\}}{}^{a}\circ^{b}{}_{S_2\sqcup\{v\}} \circ^{uv} \tag{7.29}$$

which, by (6.82) and (7.22), equals

$$-\sum_{S=S_1\sqcup S_2} \bullet_{ab}\, {}_u\bullet_v(\iota \otimes \iota)\ {}_{S_1 \sqcup \{a\}}{}^{u}\circ^{v}{}_{S_2\sqcup\{b\}} \circ^{ab}.$$

The substitution $a \leftrightarrow x$, $b \leftrightarrow y$ converts the above expression to

$$-\sum_{S=S_1\sqcup S_2} \bullet_{uv}\ {}_{a}\bullet_{b}(\iota\otimes\iota)\ {}_{S_1\sqcup\{u\}}\overset{a\ b}{\circ}{}_{S_2\sqcup\{v\}}\overset{uv}{\circ}$$

which is (7.29) with the minus sign, cf. the discussion of $\boxed{1}$.

4. *Case* $v \in A$, $u \in B$. The mirror image of the previous case.

Let us move to $\boxed{4}$. There are two possibilities.

i. *Case* $a \in A_2$. Then there exist finite sets S_1, S_2, S_3 such that $A_1 = S_1$, $A_2 = S_2 \sqcup \{a\}$ and $B = S_3$. One then rewrites $\boxed{4}$ as

$$\sum_{S_1\sqcup S_2\sqcup S_3=S} {}_{a}\bullet_{b}({}_{x}\bullet_{y}\otimes 1\!\!1)(\iota\otimes\iota\otimes\iota)({}_{S_1}\overset{x\ y}{\circ}{}_{S_2\sqcup\{a\}}\otimes 1\!\!1)\ {}_{S_1\sqcup S_2}\overset{a\ b}{\circ}{}_{S_3},$$

which, by the anti-associativity (6.80) and the coassociativity (7.20), equals

$$-\sum_{S_1\sqcup S_2\sqcup S_3=S} {}_{x}\bullet_{y}(1\!\!1\otimes {}_{a}\bullet_{b})(\iota\otimes\iota\otimes\iota)(1\!\!1\otimes {}_{S_2\sqcup\{y\}}\overset{a\ b}{\circ}{}_{S_3})\ {}_{S_1}\overset{x\ y}{\circ}{}_{S_2\sqcup S_3}. \tag{7.30}$$

ii. *Case* $a \in A_1$. There exist finite sets S_1, S_2, S_3 such that $A_1 = S_1 \sqcup \{a\}$, $A_2 = S_2$ and $B = S_3$. Expression $\boxed{4}$ then equals

$$\sum_{S_1\sqcup S_2\sqcup S_3=S} {}_{a}\bullet_{b}({}_{x}\bullet_{y}\otimes 1\!\!1)(\iota\otimes\iota\otimes\iota)({}_{S_1\sqcup\{a\}}\overset{x\ y}{\circ}{}_{S_2}\otimes 1\!\!1)\ {}_{S_1\sqcup S_2}\overset{a\ b}{\circ}{}_{S_3},$$

which we rewrite, using the symmetries (6.54) and (7.19), as

$$\sum_{S_1\sqcup S_2\sqcup S_3=S} {}_{a}\bullet_{b}({}_{y}\bullet_{x}\otimes 1\!\!1)(\iota\otimes\iota\otimes\iota)({}_{S_2}\overset{y\ x}{\circ}{}_{S_1\sqcup\{a\}}\otimes 1\!\!1)\ {}_{S_1\sqcup S_2}\overset{a\ b}{\circ}{}_{S_3}.$$

The anti-associativity (6.80) and the coassociativity (7.20) give

$$-\sum_{S_1\sqcup S_2\sqcup S_3=S} {}_{y}\bullet_{x}(1\!\!1\otimes {}_{a}\bullet_{b})(\iota\otimes\iota\otimes\iota)(1\!\!1\otimes {}_{S_2\sqcup\{x\}}\overset{a\ b}{\circ}{}_{S_3})\ {}_{S_1}\overset{y\ x}{\circ}{}_{S_2\sqcup S_3},$$

which, after the substitution $x \leftrightarrow y$, becomes (7.30). We therefore see that

$$\boxed{4} = -2\sum_{S_1\sqcup S_2\sqcup S_3=S} {}_{x}\bullet_{y}(1\!\!1\otimes {}_{a}\bullet_{b})(\iota\otimes\iota\otimes\iota)(1\!\!1\otimes {}_{S_2\sqcup\{y\}}\overset{a\ b}{\circ}{}_{S_3})\ {}_{S_1}\overset{x\ y}{\circ}{}_{S_2\sqcup S_3}.$$

The summation in $\boxed{6}$ splits into two cases, either $b \in B_1$ or $b \in B_2$. A similar analysis as the one applied to $\boxed{4}$ shows that

$$\boxed{6} = 2 \sum_{S_1 \sqcup S_2 \sqcup S_3 = S} {}_x\bullet_y (\mathbb{1} \otimes {}_a\bullet_b)(\iota \otimes \iota \otimes \iota)\big(\mathbb{1} \otimes {}_{S_2 \sqcup \{y\}}{}^{a}\circ^{b}{}_{S_3}\big)\, {}_{S_1}{}^{x}\circ^{y}{}_{S_2 \sqcup S_3},$$

which is $\boxed{4}$ with the opposite sign. This proves that $\partial^2 = 0$

Remark 7.5 Notice that the calculations in the proof of Theorem 7.2 are extremely sign-sensitive so that, e.g. the required equality $\partial^2 = 0$ in fact *determines* the signs in (6.80)–(6.83). So, even if we do not know a priory what odd modular operads are, the proof would lead us to their definition.

Let $\mathscr{T} = (\mathscr{T}, d_{\mathscr{T}})$ be an odd modular dg-operad with structure operations ${}_a\bullet_b$ and $\bullet_{uv}$, and $\mathscr{C}$ a modular cooperad with structure operations (7.16) and (7.17). The following proposition characterizes morphisms $\alpha : \mathscr{F}(\mathscr{C}) \to \mathscr{T}$. Taking as $\mathscr{T}$ the odd endomorphism operad $\mathscr{E}nd_V$ of Example 6.31, we get a description of algebras over the Feynman transform, see also [1].

Proposition 7.2 *A morphisms $\alpha : \mathscr{F}(\mathscr{C}) \to \mathscr{T}$ of odd modular dg-operads is the same as a family*

$$A = \big\{A(S; g) : \mathscr{C}(S; g) \to \mathscr{T}(S; g) \;\big|\; (S, g) \in \mathtt{Cor} \times \mathbb{A}\big\} \tag{7.31}$$

of degree 0 linear maps such that

$$A(T; g) \circ \mathscr{C}(\rho^{-1}) = \mathscr{T}(\rho) \circ A(S; g) \tag{7.32}$$

for any $g \in \mathbb{A}$ and a bijection $\rho : S \xrightarrow{\cong} T$, and such that the equality

$$d_{\mathscr{T}} A(S; g) = A(S; g) d_{\mathscr{C}} + \sum_{g'+s=g} \bullet_{uv} A(S \sqcup \{u, v\}; g') \circ^{uv}_{g} \tag{7.33}$$

$$+ \frac{1}{2} \sum_{S_1 \sqcup S_2 = S} \sum_{g_1 + g_2 = g} {}_a\bullet_b \big(A(S_1 \sqcup \{a\}; g_1) \otimes A(S_2 \sqcup \{b\}; g_2)\big)\, {}_{S_1; g_1}{}^{a}\circ^{b}{}_{S_2; g_2}$$

of maps $\mathscr{C}(S; g) \to \mathscr{T}(S; g)$ holds for all $(S, g) \in \mathtt{Cor} \times \mathbb{A}$.

Proof (of Proposition 7.1) By definition, a morphism of odd dg-modular operads

$$\alpha : \mathscr{F}(\mathscr{C}) = \big(\widetilde{\mathbb{F}}(\mathring{\mathscr{C}}), \partial + d\big) \to \mathscr{T} = (\mathscr{T}, d_{\mathscr{T}}) \tag{7.34}$$

is a morphism $\widetilde{\mathbb{F}}(\mathring{\mathscr{C}}) \to \mathscr{T}$ of odd modular operads commuting with the differentials. Since $\widetilde{\mathbb{F}}(\mathring{\mathscr{C}})$ is free, such a morphism is determined by its restriction to the module of generators, which is a morphism $A : \mathring{\mathscr{C}} \to \mathscr{T}$ of modular modules. By the definition of the modular module $\mathring{\mathscr{C}}$ associated with the module comodule $\mathscr{C}$, A is precisely the family (7.31) satisfying (7.32).

It remains to express the condition that (7.34) commutes with the differentials, i.e. that

$$d_{\mathscr{T}}\alpha(S; g) = \alpha(S; g)(\partial + d) \tag{7.35}$$

for each $S \in \mathtt{Cor}$ and $g \in \mathbb{A}$, in terms of the family A. Since a composition of a derivation with a morphism is again a derivation, it is enough by Theorem 7.1 to verify (7.35) on the generators $\mathring{\mathscr{C}}$. Let us apply $\alpha(S; g)$ to the first term in the right-hand side of (7.28) defining ∂ on $\mathring{\mathscr{C}}(S; g)$. Using the fact that α extends A, we see that

$$\alpha(S; g) \sum_{g'+s=g} \bullet_{uv}\, \iota \circ_{g'}^{uv} = \sum_{g'+s=g} \bullet_{uv} \alpha(S \sqcup \{uv\}; g')\iota\circ_{g'}^{uv} = \sum_{g'+s=g} \bullet_{uv}\, A(S \sqcup \{uv\}; g')\circ_{g'}^{uv}$$

Applying $\alpha(S; g)$ to the second term in the right-hand side of (7.28) gives

$$\alpha(S; g) \sum_{S_1 \sqcup S_2 = S}\ \sum_{g_1+g_2=g} {}_a\bullet_b(\iota \otimes \iota)\ {}_{S_1;\,g_1}{\overset{a\ b}{\circ}}_{S_2;g_2}$$

$$= \sum_{S_1 \sqcup S_2 = S}\ \sum_{g_1+g_2=g} {}_a\bullet_b \big(\alpha(S_1 \sqcup \{a\}; g_1) \otimes \alpha(S_2 \sqcup \{b\}; g_2)\big)(\iota \otimes \iota)\ {}_{S_1;\,g_1}{\overset{a\ b}{\circ}}_{S_2;g_2}$$

$$= \sum_{S_1 \sqcup S_2 = S}\ \sum_{g_1+g_2=g} {}_a\bullet_b \circ\big(A(S_1 \sqcup \{a\}; g_1) \otimes A(S_2 \sqcup \{b\}; g_2)\big) \circ\ {}_{S_1;\,g_1}{\overset{a\ b}{\circ}}_{S_2;g_2}.$$

Finally, $\alpha(S; g)d$ restricted to $\mathring{\mathscr{C}}$ clearly equals $A(S; g)d_{\mathscr{C}}$, so the right-hand side of (7.33) equals $\alpha(S; g)(\partial + d)$ restricted to $\mathring{\mathscr{C}}$. The fact that $d_{\mathscr{T}}A(S; g)$ is $d_{\mathscr{T}}\alpha(S; g)$ restricted to $\mathscr{C}(S; g)$ finishes the proof.

Remark 7.6 Theorem 8.2 below will use a skeletal version of Proposition 7.2. As before, $[n] := \{1, \dots, n\}$ for $n \in \mathbb{N}$, and denote $\mathscr{C}(n; g) := \mathscr{C}\big([n]; g\big)$ and $\mathscr{T}(n; g) := \mathscr{T}\big([n]; g\big)$, $g \in \mathbb{A}$. Each $\mathscr{T}(n; g)$ is a natural left Σ_n-module, while each $\mathscr{C}(n; g)$ is a natural right Σ_n-module. We claim that the family (7.31) is determined by the sequence

$$A_{\mathrm{sk}} = \big\{A(n; g) : \mathscr{C}(n; g) \to \mathscr{T}(n; g) \mid (S, g) \in \mathbb{N} \times \mathbb{A}\big\} \tag{7.36}$$

of linear degree 0 maps such that

$$A(n; g) = \mathscr{T}(\sigma) \circ A(n; g) \circ \mathscr{C}(\sigma) \tag{7.37}$$

for each $n \in \mathbb{N}$, $g \in \mathbb{A}$ and a permutation $\sigma : [n] \xrightarrow{\cong} [n] \in \Sigma_n$. In elements, Eq. (7.37) means that

$$A(n; g)(c) = \sigma\, A(n; g)(c\sigma)$$

for each $c \in \mathscr{C}(n; g)$. Let us verify this statement.

Each family (7.31) determines the skeletal family (7.36) by restricting to finite sets of the form $[n]$, $n \in \mathbb{N}$. Condition (7.37) for this restricted family follows from (7.32) taken with $S = T = [n]$ and $\rho = \sigma : [n] \xrightarrow{\cong} [n] \in \Sigma_n$.

On the other hand, suppose that we are given a skeletal family $A_{\rm sk}$ as in (7.36). For a finite set S choose an isomorphism $\rho : [n] \xrightarrow{\cong} S$ and define $A(S; g)$ in (7.31) by

$$A(S; g) := \mathscr{T}(\rho) \circ A(n; g) \circ \mathscr{C}(\rho). \tag{7.38}$$

It follows from the equivariance (7.37) that $A(S; g)$ does not depend on the concrete choice of ρ and that the family defined this way satisfies (7.32). The correspondence $A \leftrightarrow A_{\rm sk}$ described above is clearly one-to-one.

Equation (7.33) with $S = [n]$ reads

$$d_{\mathscr{T}} A(n; g) = A(n; g) d_{\mathscr{C}} + \sum_{g'+s=g} \bullet_{uv} A\big([n] \sqcup \{u, v\}; g'\big) \circ^{uv}_{g'}$$
$$+ \frac{1}{2} \sum_{S_1 \sqcup S_2 = [n]} \sum_{g_1+g_2=g} {}_a\bullet_b \big(A(S_1 \sqcup \{a\}; g_1) \otimes A(S_2 \sqcup \{b\}; g_2)\big)\, {}_{S_1;\, g_1}{}^{a}\circ^{b}{}_{S_2; g_2}$$

Expressing $A\big([n] \sqcup \{u, v\}; g'\big)$, $A\big(S_1 \sqcup \{a\}; g_1\big)$ and $A\big(S_2 \sqcup \{b\}; g_2\big)$ via the skeletal family $A_{\rm sk}$ using (7.38) converts this equation into

$$d_{\mathscr{T}} A(n; g) = A(n; g) d_{\mathscr{C}} + \sum_{g'+s=g} \bullet_{uv} \mathscr{T}(\theta) A(n+2; g') \mathscr{C}(\theta) \circ^{uv}_{g'} \tag{7.39}$$
$$+ \frac{1}{2} \sum_{S_1 \sqcup S_2 = [n]} \sum_{g_1+g_2=g} {}_a\bullet_b (\theta_1 \otimes \theta_2)\big(A(n_1{+}1; g_1) \otimes A(n_2{+}1; g_2)\big)(\theta_1 \otimes \theta_2)\, {}_{S_1;\, g_1}{}^{a}\circ^{b}{}_{S_2; g_2}$$

for chosen isomorphisms $\theta : [n+2] \xrightarrow{\cong} [n] \sqcup \{u, v\}$, $\theta_1 : [n_1 + 1] \xrightarrow{\cong} S_1 \sqcup \{a\}$ and $\theta_2 : [n_2 + 1] \xrightarrow{\cong} S_2 \sqcup \{b\}$. To fit the above formula into a display of finite width, we wrote in the second line

$$(\theta_1 \otimes \theta_2)\big(A(n_1 + 1; g_1) \otimes A(n_2 + 1; g_2)\big)(\theta_1 \otimes \theta_2)$$

instead of

$$\big(\mathscr{T}(\theta_1) \otimes \mathscr{T}(\theta_2)\big)\big(A(n_1+1; g_1) \otimes A(n_2+1; g_2)\big)\big(\mathscr{C}(\theta_1) \otimes \mathscr{C}(\theta_2)\big).$$

References

1. Barannikov, S.: Modular operads and Batalin-Vilkovisky geometry. Int. Math. Res. Not. **2007**(19), 31 (2007). Art. ID rnm075. http://dx.doi.org/10.1093/imrn/rnm075
2. Getzler, E., Kapranov, M.: Modular operads. Compos. Math. **110**(1), 65–126 (1998)
3. Markl, M., Shnider, S., Stasheff, J.: Operads in algebra, topology and physics. In: Mathematical Surveys and Monographs, vol. 96. American Mathematical Society, Providence, RI (2002)

Structures Relevant to Physics

8

The last chapter the book is devoted to the mathematical interpretation of physical objects discussed in Part I. The standard references are [1, 4, 9] and [12].

8.1 BV Algebras and the Master Equation

Generalizing Barannikov's [1], we prove that an odd dg-modular operad morphism $\alpha : \mathscr{F}(\mathscr{C}) \to \mathscr{T}$ as in Proposition 7.2 can be, in the case when $\mathscr{C}$ is a component-wise linear dual of a modular operad $\mathscr{M}$ of finite type as in Proposition 7.1, conveniently described via a solution of a certain master equation in a shifted dg-Lie algebra succinctly defined in terms of $\mathscr{M}$ and $\mathscr{T}$. Recall that $[n]$ for $n \geq 0$ denotes the set $\{1, \dots, n\}$, with $[0]$ interpreted as the empty set. We denote, as usual, $\mathscr{M}(n; g) := \mathscr{M}\big([n]; g\big)$ and $\mathscr{T}(n; g) := \mathscr{T}\big([n]; g\big)$, $g \in \mathbb{A}$.

Definition 8.1 Let $\mathscr{M} = (\mathscr{M}, d_{\mathscr{M}})$ be a dg-modular operad with structure operations ${}_a\circ_b$ and $\circ_{uv}$, and $\mathscr{T} = (\mathscr{T}, d_{\mathscr{T}})$ an odd dg-modular operad with structure operations ${}_a\bullet_b$ and $\bullet_{uv}$. Define

$$MT(n; g) := \big(\mathscr{M}(n; g) \otimes \mathscr{T}(n; g)\big)^{\Sigma_n}$$

to be, for $n \in \mathbb{N}$, $g \in \mathbb{A}$, the space of invariants under the diagonal action of the symmetric group Σ_n on the tensor product $\mathscr{M}(n; g) \otimes \mathscr{T}(n; g)$.

Let us introduce the following three operations. For $e \in MT(n; g)$ put

$$d(e) := \big(d_{\mathscr{M}} \otimes \mathbb{1}_{\mathscr{T}(n;g)} - \mathbb{1}_{\mathscr{M}(n;g)} \otimes d_{\mathscr{T}}\big)(e) \in MT(n; g). \tag{8.1}$$

© Springer Nature Switzerland AG 2020

M. Doubek et al., *Algebraic Structure of String Field Theory*, Lecture Notes in Physics 973, https://doi.org/10.1007/978-3-030-53056-3_8

For $f \in MT(n+2; g)$, let

$$\Delta(f) := \big(\circ_{uv} \otimes \bullet_{uv}\big)\big(\mathscr{M}(\theta) \otimes \mathscr{T}(\theta)\big)(f) \in MT(n; g+s) \tag{8.2}$$

for an arbitrary bijection $\theta : [n+2] \xrightarrow{\cong} [n] \sqcup \{u, v\}$. Finally, for $g \in MT(n_1+1; g_1)$ and $h \in MT(n_2+1; g_2)$, let $\{g, h\} \in MT(n_1+n_2; g_1+g_2)$ be defined as

$$\{g, h\} := \sum_{S_1 \sqcup S_2 = [n_1+n_2]} ({}_a\circ_b \otimes {}_a\bullet_b)\tau\big(\mathscr{M}(\theta_1) \otimes \mathscr{T}(\theta_1) \otimes \mathscr{M}(\theta_2) \otimes \mathscr{T}(\theta_2)\big)(g \otimes h), \tag{8.3}$$

where $\theta_1 : [n_1+1] \xrightarrow{\cong} S_1 \sqcup \{a\}$ and $\theta_2 : [n_2+1] \xrightarrow{\cong} S_2 \sqcup \{b\}$ are arbitrary bijections[1] and τ exchanges the two middle factors. We complete the definition by setting $\Delta(f) := 0$ for $f \in MT(n; g)$, $n = 0, 1$, and $\{g, h\} := 0$ if $g \in MT(0; g_1)$ or $h \in MT(0; g_2)$.

Lemma 8.1 *The above operations are well defined and do not depend on the choices of θ in (8.2) and θ_1, θ_2 in (8.3).*

Proof. It follows from the Σ_n-equivariance of the differentials $d_{\mathscr{T}}$ and $d_{\mathscr{M}}$ that $d(e)$ in (8.1) is Σ_n-stable, too, i.e. indeed $d(e) \in MT(n; g)$. Let us prove that $\Delta(f)$ as defined in (8.2) does not depend on θ.

Assume that $\theta' : [n+2] \to [n] \sqcup \{u, v\}$ is another bijection. Then

$$\begin{aligned}\big(\mathscr{M}(\theta') \otimes \mathscr{T}(\theta')\big)(f) &= \big(\mathscr{M}(\theta) \otimes \mathscr{T}(\theta)\big)\big(\mathscr{M}(\theta^{-1}\theta') \otimes \mathscr{T}(\theta^{-1}\theta')\big)(f)\\ &= \big(\mathscr{M}(\theta) \otimes \mathscr{T}(\theta)\big)(f),\end{aligned}$$

since $\theta^{-1}\theta' \in \Sigma_{n+2}$ and $f \in \mathscr{M}([n+2]; g) \otimes \mathscr{T}([n+2]; g)$ is Σ_{n+2}-invariant by assumption.

Let us prove that $\Delta(f) \in \mathscr{M}([n]; g+s) \otimes \mathscr{T}([n]; g+s)$ is Σ_n-invariant. Invoking (6.53) resp. its obvious version for odd modular operads, we get for $\sigma \in \Sigma_n$

$$\begin{aligned}\sigma\Delta(f) &= \big(\mathscr{M}(\sigma) \otimes \mathscr{T}(\sigma)\big)(\circ_{uv} \otimes \bullet_{uv})\big(\mathscr{M}(\theta) \otimes \mathscr{T}(\theta)\big)(f)\\ &= (\circ_{uv} \otimes \bullet_{uv})\big(\mathscr{M}(\tilde{\sigma}\theta) \otimes \mathscr{T}(\tilde{\sigma}\theta)\big)(f),\end{aligned}$$

where $\tilde{\sigma} : [n] \sqcup \{u, v\} \xrightarrow{\cong} [n] \sqcup \{u, v\}$ fixes u, v and restricts to σ on $[n]$. Since $\tilde{\sigma}\theta$ is just another isomorphism between $[n+2]$ and $[n] \sqcup \{u, v\}$, $\sigma\Delta(f)$ equals $\Delta(f)$ by the previous paragraph.

The independence of $\{g, h\}$ in (8.3) on θ_1 and θ_2 is proved precisely as the independence of $\Delta(f)$ on θ. Let us show that $\{g, h\}$ is $\Sigma_{n_1+n_2}$-invariant. Since

[1] Notice that necessarily $|S_1| = n_1$ and $|S_2| = n_2$.

$[n_1+n_2] = S_1 \sqcup S_2$, a permutation $\sigma \in \Sigma_{n_1+n_2}$ is the same as an isomorphism $S_1 \sqcup S_2 \xrightarrow{\cong} S_1 \sqcup S_2$. Using (6.53) resp. its odd version we see that, for such a σ,

$$\big(\mathscr{M}(\sigma)\otimes\mathscr{T}(\sigma)\big)\big({}_a\circ_b \otimes {}_a\bullet_b\big) = \big({}_a\circ_b \otimes {}_a\bullet_b\big)\big(\mathscr{M}(\tilde\sigma_1)\otimes\mathscr{M}(\tilde\sigma_2)\otimes\mathscr{T}(\tilde\sigma_1)\otimes\mathscr{T}(\tilde\sigma_2)\big),$$

where $\tilde\sigma_1 : S_1 \sqcup \{a\} \xrightarrow{\cong} \sigma(S_1) \sqcup \{a\}$ is the isomorphism that restricts to σ on S_1, and $\tilde\sigma_2 : S_2 \sqcup \{b\} \xrightarrow{\cong} \sigma(S_2) \sqcup \{b\}$ is defined analogously. The above equality implies that

$$\begin{aligned}&\sigma\{g,h\}\\ &= \sum_{S_1\sqcup S_2=[n_1+n_2]} \big({}_a\circ_b \otimes {}_a\bullet_b\big)\tau\big(\mathscr{M}(\tilde\sigma_1\theta_1)\otimes\mathscr{T}(\tilde\sigma_1\theta_1)\otimes\mathscr{M}(\tilde\sigma_2\theta_2)\otimes\mathscr{T}(\tilde\sigma_2\theta_2)\big)(g\otimes h).\end{aligned}$$

Replacing the summation over S_1, S_2 by the summation over $\sigma(S_1), \sigma(S_2)$, and the corresponding isomorphisms θ_1, θ_2 by $\tilde\sigma_1\theta_1, \tilde\sigma_1\theta_2$ we conclude that the last expression equals $\{g,h\}$ as desired.

Let us introduce the total graded vector space

$$MT := \prod_{n\geq 0,\ g\in\mathbb{A}} MT(n;g). \tag{8.4}$$

The operations defined in (8.1)–(8.3) act on sequences in MT, defining degree $+1$ operations $d, \Delta : MT \to MT$ and $\{-,-\} : MT \otimes MT \to MT$.

Theorem 8.1 *The object* $MT = \big(MT, d, \Delta, \{-,-\}\big)$ *is a desuspended bi-differential graded Lie algebra, i.e. the operations* $d : MT \to MT$, $\Delta : MT \to MT$ *and* $\{-,-\} : MT \otimes MT \to MT$ *have degree* $+1$, $\{-,-\}$ *is graded symmetric and the following axioms are fulfilled:*[2]

$$d^2 = \Delta d + d\Delta = \Delta^2 = 0, \tag{8.5}$$

$$d\{f,g\} + \{df,g\} + (-1)^{|f|}\{f,dg\} = 0, \tag{8.6}$$

$$\Delta\{f,g\} + \big\{\Delta(f),g\big\} + (-1)^{|f|}\big\{f,\Delta(g)\big\} = 0, \text{ and} \tag{8.7}$$

$$(-1)^{|f||h|}\big\{\{f,g\},h\big\} + (-1)^{|h||g|}\big\{\{h,f\},g\big\} + (-1)^{|g||f|}\big\{\{g,h\},f\big\} = 0 \tag{8.8}$$

for arbitrary homogeneous $f, g, h \in MT$.

[2]Equivalently, the operations d, Δ and $\{-,-\}$ induce on the suspension $\uparrow MT$ a dg-Lie algebra structure.

Proof. The vanishing of d^2 is obvious, and also the equation $\Delta d + d\Delta = 0$ readily follows from the fact that the differentials $d_{\mathscr{M}}$ resp. $d_{\mathscr{T}}$ commute with the contractions $\circ_{uv}$ resp. $\bullet_{uv}$. Let us prove $\Delta^2 = 0$.

It is easy to verify that for $f \in MT(n; g)$,

$$\Delta^2(f) = \big(\circ_{ab} \circ_{cd} \otimes \bullet_{ab} \bullet_{cd} \big)\big(\mathscr{M}(\theta) \otimes \mathscr{T}(\theta)\big)(f), \tag{8.9}$$

where $\theta : [n] \xrightarrow{\cong} [n-4] \sqcup \{a, b, c, d\}$ is an arbitrary bijection.[3] Now consider the isomorphism

$$\sigma : [n-4] \sqcup \{a, b, c, d\} \xrightarrow{\cong} [n-4] \sqcup \{a, b, c, d\}$$

with

$$\sigma(a) := c,\ \sigma(b) := d,\ \sigma(c) := a \text{ and } \sigma(d) := b$$

that restricts to the identity on $[n-4]$. Since $\sigma\theta$ is just another isomorphism between $[n]$ and $[n-4] \sqcup \{a, b, c, d\}$, $\Delta^2(f)$ in (8.9) equals

$$\big(\circ_{ab} \circ_{cd} \otimes \bullet_{ab} \bullet_{cd} \big)\big(\mathscr{M}(\sigma\theta) \otimes \mathscr{M}(\sigma\theta)\big)(f).$$

By the equivariance of the contractions, the expression in the above display equals

$$\big(\circ_{cd} \circ_{ab} \otimes \bullet_{cd} \bullet_{ab} \big)\big(\mathscr{M}(\theta) \otimes \mathscr{M}(\theta)\big)(f)$$

while, by the commutativity (6.56) resp. the anti-commutativity (6.81), this equals

$$-\big(\circ_{ab} \circ_{cd} \otimes \bullet_{ab} \bullet_{cd} \big)\big(\mathscr{M}(\theta) \otimes \mathscr{M}(\theta)\big)(f),$$

which we recognize as the right-hand side of (8.9) with the minus sign. We conclude that $\Delta^2(f) = 0$. This finishes the verification of (8.5).

Equation (8.6) follows from the fact that both differentials $d_{\mathscr{M}}$ and $d_{\mathscr{T}}$ are derivations with respect to the structure operations. In the rest of this section we shorten the formulas by denoting the actions as e.g. $\mathscr{M}(\theta)$ resp. $\mathscr{T}(\theta)$ by θ, the exact meaning will always be clear from the context. For instance, formula (8.9) will read

$$\Delta^2(f) = \big(\circ_{ab} \circ_{cd} \otimes \bullet_{ab} \bullet_{cd} \big)(\theta \otimes \theta)(f).$$

[3] We tacitly assume here that $n \geq 4$. When $n < 4$, $\Delta^2(f) = 0$ immediately from definition. We use this kind of assumptions throughout the rest of the proof.

Let us verify (8.7). A tedious but straightforward calculation yields that, for $f \in MT(n_1; g_1)$ and $g \in MT(n_2; g_2)$,

$$\Delta\{f, g\} = \left(\boxed{1} + 2\boxed{2} + \boxed{3}\right)(f \otimes g), \tag{8.10}$$

where

$$\boxed{1} := \sum_{S_1 \sqcup S_2 = [n_1+n_2-4]} \left(\circ_{uv}\ {}_a\circ_b \otimes \bullet_{uv}\ {}_a\bullet_b \right)\tau(\theta'_1 \otimes \theta'_1 \otimes \theta'_2 \otimes \theta'_2)$$

$$\boxed{2} := \sum_{S_1 \sqcup S_2 = [n_1+n_2-4]} \left(\circ_{uv}\ {}_a\circ_b \otimes \bullet_{uv}\ {}_a\bullet_b \right)\tau(\theta_1 \otimes \theta_1 \otimes \theta_2 \otimes \theta_2)$$

$$\boxed{3} := \sum_{S_1 \sqcup S_2 = [n_1+n_2-4]} \left(\circ_{uv}\ {}_a\circ_b \otimes \bullet_{uv}\ {}_a\bullet_b \right)\tau(\theta''_1 \otimes \theta''_1 \otimes \theta''_2 \otimes \theta''_2)$$

where the bijections

$$\theta'_1 : [n_1] \xrightarrow{\cong} S_1 \sqcup \{a, u, v\},\ \theta'_2 : [n_2] \xrightarrow{\cong} S_2 \sqcup \{b\} \text{ in } \boxed{1},$$
$$\theta_1 : [n_1] \xrightarrow{\cong} S_1 \sqcup \{a, u\},\ \theta_2 : [n_2] \xrightarrow{\cong} S_2 \sqcup \{b, v\} \text{ in } \boxed{2}, \text{ and}$$
$$\theta''_1 : [n_1] \xrightarrow{\cong} S_1 \sqcup \{a\},\ \theta''_2 : [n_2] \xrightarrow{\cong} S_2 \sqcup \{b, u, v\} \text{ in } \boxed{3},$$

are arbitrary. The terms in the three sums are symbolized respectively as

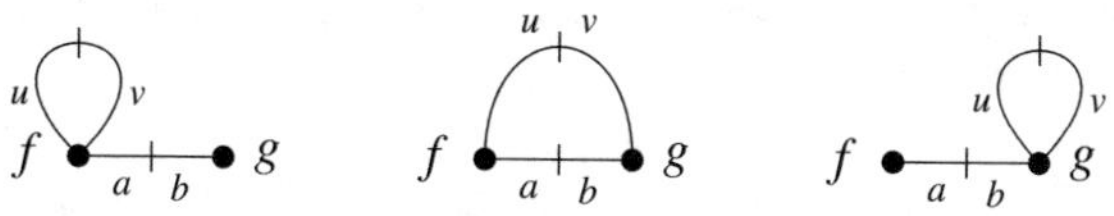

Let us show that $\boxed{2}$ vanishes. To this end, recall that the expression

$$\left(\circ_{uv}\ {}_a\circ_b \otimes \bullet_{uv}\ {}_a\bullet_b \right)\tau(\theta_1 \otimes \theta_1 \otimes \theta_2 \otimes \theta_2) \tag{8.11}$$

in the sum does not depend on the particular choices of θ_1 and θ_2. Precomposing θ_1 with the isomorphism $\sigma_1 : S_1 \sqcup \{a, u\} \xrightarrow{\cong} S_1 \sqcup \{a, u\}$ that interchanges a with u and restricts to the identity on S_1, and θ_2 with the similar isomorphism σ_2 interchanging b with v thus does not change the value of (8.11) which therefore equals

$$\left(\circ_{uv}\ {}_a\circ_b \otimes \bullet_{uv}\ {}_a\bullet_b \right)\tau(\sigma_1\theta_1 \otimes \sigma_1\theta_1 \otimes \sigma_2\theta_2 \otimes \sigma_2\theta_2)$$

which, by the equivariance of the structure operations of $\mathscr{M}$ and $\mathscr{T}$, equals

$$\left(\circ_{ab}\ {}_u\circ_v \otimes \bullet_{ab}\ {}_u\bullet_v \right)\tau(\theta_1 \otimes \theta_1 \otimes \theta_2 \otimes \theta_2)$$

which, in turn, equals

$$-\left(\circ_{uv}\ {}_a\circ_b \otimes\ \bullet_{uv}\ {}_a\bullet_b \right)\tau(\theta_1 \otimes \theta_1 \otimes \theta_2 \otimes \theta_2)$$

by the commutativity (6.57) resp. the anti-commutativity (6.82). Comparing it with (8.11) we conclude that the middle term $\boxed{2}$ of (8.10) vanishes.

A straightforward calculation shows that, for f, g as in (8.10),

$$\left\{\Delta(f), g\right\} = \sum_{S_1 \sqcup S_2 = [n_1+n_2-4]} \left({}_a\circ_b(\circ_{uv} \otimes 1\!\!1) \otimes {}_a\bullet_b(\bullet_{uv} \otimes 1\!\!1)\right)\tau(\theta_1' \otimes \theta_1' \otimes \theta_2' \otimes \theta_2')(f \otimes g)$$

for some bijections $\theta_1' : [n_1] \to S_1 \sqcup \{a, u, v\}$ and $\theta_2' : [n_2] \to S_2 \sqcup \{b\}$. The sum in the right-hand side however equals

$$\sum_{S_1 \sqcup S_2 = [n_1+n_2-4]} -\left(\circ_{uv}\ {}_a\circ_b \otimes\ \bullet_{uv}\ {}_a\bullet_b \right)\tau(\theta_1' \otimes \theta_1' \otimes \theta_2' \otimes \theta_2')(f \otimes g)$$

by the commutativity (6.58) resp. the anti-commutativity (6.83), which is $\boxed{1}(f \otimes g)$ with the minus sign. By exactly the same method we show that

$$(-1)^{|f|}\left\{f, \Delta(g)\right\} = -\boxed{3}(f \otimes g).$$

This finishes the proof of (8.7).

Let us move to the proof of the Jacobi identity (8.8). It will be convenient to introduce two auxiliary maps. For finite sets S_1, S_2, S_3, integers $p, q, r \geq 0$, genera $i, j, k \in \mathbb{A}$ and isomorphisms

$$\theta_1 : [p] \xrightarrow{\cong} S_1 \sqcup \{b\}, \theta_2 : [q] \xrightarrow{\cong} S_2 \sqcup \{a, c\}, \theta_3 : [r] \xrightarrow{\cong} S_3 \sqcup \{d\}$$

the first map

$$A(\theta_1, \theta_2, \theta_3) : MT(p; i) \otimes MT(q; j) \otimes MT(r; k) \to$$
$$\to \mathscr{M}(S_1 \sqcup S_2 \sqcup S_3; i + j + k) \otimes \mathscr{T}(S_1 \sqcup S_2 \sqcup S_3; i + j + k)$$

is defined by

$$A(\theta_1, \theta_2, \theta_3) := \left({}_c\circ_d({}_b\circ_a \otimes 1\!\!1)(\theta_1 \otimes \theta_2 \otimes \theta_3) \otimes {}_c\bullet_d({}_b\bullet_a \otimes 1\!\!1)(\theta_1 \otimes \theta_2 \otimes \theta_3)\right)\psi,$$

where the isomorphism ψ interchanges the tensor factors of the subspace

$$MT(p; i) \otimes MT(q; j) \otimes MT(r; k)$$

of the tensor product

$$\mathcal{M}(p;i)\otimes\mathcal{T}(p;i)\otimes\mathcal{M}(q;j)\otimes\mathcal{T}(q;j)\otimes\mathcal{M}(r;k)\otimes\mathcal{T}(r;k)$$

according to the permutation

$$(1,2,3,4,5,6)\longmapsto(1,4,2,5,3,6).$$

The second map

$$B(\theta_2,\theta_3,\theta_1): MT(q;j)\otimes MT(r;k)\otimes MT(p;i)\to$$
$$\to\mathcal{M}(S_1\sqcup S_2\sqcup S_3;i+j+k)\otimes\mathcal{T}(S_1\sqcup S_2\sqcup S_3;i+j+k)$$

is given by the formula

$$B(\theta_2,\theta_3,\theta_1):=\big({}_a\circ_b({}_c\circ_d\otimes\mathbb{1})(\theta_2\otimes\theta_3\otimes\theta_1)\otimes{}_a\bullet_b({}_c\bullet_d\otimes\mathbb{1})(\theta_2\otimes\theta_3\otimes\theta_1)\big)\psi.$$

The nature of $A(\theta_1,\theta_2,\theta_3)(x\otimes y\otimes z)$ and $B(\theta_2,\theta_3,\theta_1)(y\otimes z\otimes x)$ is, for

$$x\in MT(p;i),\ y\in MT(q;j)\text{ and }z\in MT(r;k)$$

symbolized respectively as

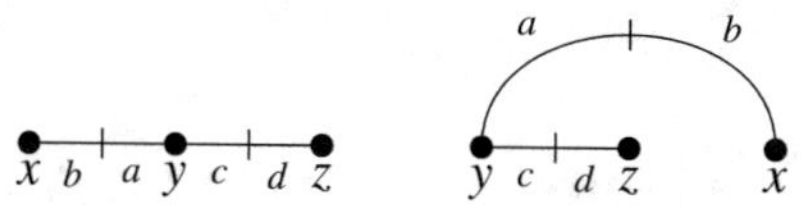

Denote by

$$\lambda: MT(p;i)\otimes MT(q;j)\otimes MT(r;k)\xrightarrow{\cong}MT(q;j)\otimes MT(r;k)\otimes MT(p;i)\qquad(8.12)$$

the isomorphism that permutes the tensor factors according to the permutation $(1,2,3)\mapsto(2,3,1)$. The crucial property of the auxiliary maps is that

$$A(\theta_1,\theta_2,\theta_3)=-B(\theta_2,\theta_3,\theta_1)\lambda.\qquad(8.13)$$

Indeed, $A(\theta_1,\theta_2,\theta_3)$ which, by definition, equals

$$\big({}_c\circ_d({}_a\circ_b\otimes\mathbb{1})(\theta_1\otimes\theta_2\otimes\theta_3)\otimes{}_c\bullet_d({}_a\bullet_b\otimes\mathbb{1})(\theta_1\otimes\theta_2\otimes\theta_3)\big)\psi$$

can be, using (6.55) and (6.80), rewritten as

$$-\big({}_b\circ_a(\mathbb{1}\otimes {}_c\circ_d)(\theta_1\otimes\theta_2\otimes\theta_3)\otimes {}_b\bullet_a(\mathbb{1}\otimes {}_c\bullet_d)(\theta_1\otimes\theta_2\otimes\theta_3)\big)\psi.$$

This in turn, by (6.54) and its odd version, equals

$$-\big({}_a\circ_b({}_c\circ_d\otimes\mathbb{1})(\theta_2\otimes\theta_3\otimes\theta_1)\otimes {}_a\bullet_b({}_c\bullet_d\otimes\mathbb{1})(\theta_2\otimes\theta_3\otimes\theta_1)\big)\psi\lambda,$$

which is $-B(\theta_2,\theta_3,\theta_1)\lambda$ as claimed.

The terms in the Jacobi identity can be expressed via the auxiliary maps as follows. For $f\in MT(n_1;g_1)$, $g\in MT(n_2;g_2)$, and $h\in MT(n_3;g_3)$ one obtains

$$\big\{\{f,g\},h\big\}=\sum\big(A(\theta_{11},\theta_{22},\theta_{33})+B(\theta_{12},\theta_{23},\theta_{31})\big)(f\otimes g\otimes h),$$

where the summation running over all disjoint partitions

$$S_1\sqcup S_2\sqcup S_3=[n_1+n_2+n_3-4]$$

with arbitrarily chosen isomorphism

$$\theta_{11}:[n_1]\xrightarrow{\cong}S_1\sqcup\{b\},\ \theta_{22}:[n_2]\xrightarrow{\cong}S_2\sqcup\{a,c\},\ \theta_{33}:[n_3]\xrightarrow{\cong}S_3\sqcup\{d\}\ \text{ and}$$
$$\theta_{12}:[n_1]\xrightarrow{\cong}S_2\sqcup\{a,c\},\theta_{23}:[n_2]\xrightarrow{\cong}S_3\sqcup\{d\},\ \theta_{31}:[n_3]\xrightarrow{\cong}S_1\sqcup\{b\}.$$

Likewise, for the same f, g, and h one has

$$\big\{\{g,h\},f\big\}=\sum\big(A(\theta_{21},\theta_{32},\theta_{13})+B(\theta_{22},\theta_{33},\theta_{11})\big)(g\otimes h\otimes f),$$

with chosen isomorphism

$$\theta_{21}:[n_2]\xrightarrow{\cong}S_1\sqcup\{b\},\ \theta_{32}:[n_3]\xrightarrow{\cong}S_2\sqcup\{a,c\},\ \theta_{13}:[n_1]\xrightarrow{\cong}S_3\sqcup\{d\}\ \text{ and}$$
$$\theta_{22}:[n_2]\xrightarrow{\cong}S_2\sqcup\{a,c\},\theta_{33}:[n_3]\xrightarrow{\cong}S_3\sqcup\{d\},\ \theta_{11}:[n_1]\xrightarrow{\cong}S_1\sqcup\{b\}.$$

Finally,

$$\big\{\{h,f\},g\big\}=\sum\big(A(\theta_{31},\theta_{12},\theta_{23})+B(\theta_{32},\theta_{13},\theta_{21})\big)(h\otimes f\otimes g),$$

with isomorphism

$$\theta_{31}:[n_3]\xrightarrow{\cong}S_1\sqcup\{b\},\ \theta_{12}:[n_1]\xrightarrow{\cong}S_2\sqcup\{a,c\},\ \theta_{23}:[n_2]\xrightarrow{\cong}S_3\sqcup\{d\}\ \text{ and}$$
$$\theta_{32}:[n_3]\xrightarrow{\cong}S_2\sqcup\{a,c\},\theta_{13}:[n_1]\xrightarrow{\cong}S_3\sqcup\{d\},\ \theta_{21}:[n_2]\xrightarrow{\cong}S_1\sqcup\{b\}.$$

The Jacobi identity (8.8) multiplied by $(-1)^{|f||h|}$ is then expressed as

$$
\begin{aligned}
0 = &\sum \big(A(\theta_{11}, \theta_{22}, \theta_{33}) + B(\theta_{12}, \theta_{23}, \theta_{31})\big)(f \otimes g \otimes h)\\
&+(-1)^{|f|(|g|+|h|)} \sum \big(A(\theta_{21}, \theta_{32}, \theta_{13}) + B(\theta_{22}, \theta_{33}, \theta_{11})\big)(g \otimes h \otimes f)\\
&+(-1)^{|h|(|f|+|g|)} \sum \big(A(\theta_{31}, \theta_{12}, \theta_{23}) + B(\theta_{32}, \theta_{13}, \theta_{21})\big)(h \otimes f \otimes g)
\end{aligned}
$$

or equivalently, using the isomorphism λ of (8.12) and invoking the Koszul sign convention, as

$$
\begin{aligned}
0 = &\sum \big(A(\theta_{11}, \theta_{22}, \theta_{33}) + B(\theta_{12}, \theta_{23}, \theta_{31})\big)(f \otimes g \otimes h)\\
&+ \sum \big(A(\theta_{21}, \theta_{32}, \theta_{13})\lambda + B(\theta_{22}, \theta_{33}, \theta_{11})\lambda\big)(f \otimes g \otimes h) \qquad (8.14)\\
&+ \sum \big(A(\theta_{31}, \theta_{12}, \theta_{23})\lambda^2 + B(\theta_{32}, \theta_{13}, \theta_{21})\lambda^2\big)(f \otimes g \otimes h).
\end{aligned}
$$

Equation (8.13) readily implies that

$$
\begin{aligned}
A(\theta_{11}, \theta_{22}, \theta_{33}) &= -B(\theta_{22}, \theta_{33}, \theta_{11})\lambda\\
A(\theta_{21}, \theta_{32}, \theta_{13})\lambda &= -B(\theta_{32}, \theta_{13}, \theta_{21})\lambda^2 \text{ and}\\
A(\theta_{31}, \theta_{12}, \theta_{23})\lambda^2 &= -B(\theta_{12}, \theta_{23}, \theta_{31})
\end{aligned}
$$

from which (8.15) follows immediately. This finishes the proof of Theorem 8.1.

The operations introduced in Definition 8.1 can be expressed using the skeletal versions $\big(\{\mathscr{M}(n)\}_{n\geq 0},\ {}_a\overline{\circ}_b, \overline{\circ}_{ij}\big)$ resp. $\big(\{\mathscr{T}(n)\}_{n\geq 0},\ {}_a\overline{\bullet}_b, \overline{\bullet}_{ij}\big)$ of the modular operad $\mathscr{M}$ resp. the odd modular operad $\mathscr{T}$ as follows.

Proposition 8.1 *For $f \in MT(n+2; g)$,*

$$
\Delta(f) = \big(\overline{\circ}_{ij} \otimes \overline{\bullet}_{ij}\big)(f) \qquad (8.15)
$$

with arbitrary fixed $i, j \in [n+1]$. Let $g \in MT(n_1+1; g_1)$ and $h \in MT(n_2+1; g_2)$. Choose $i \in [n_1+1]$, $j \in [n_2+2]$ arbitrarily. Then

$$
\{g, h\} := \sum_{\rho} \big(\mathscr{M}(\rho) \otimes \mathscr{T}(\rho)\big)\big({}_i\overline{\circ}_j \otimes {}_i\overline{\bullet}_j\big)\tau(g \otimes h), \qquad (8.16)
$$

where the summation runs over all isomorphism $\rho : [n_1+n_2] \xrightarrow{\cong} [n_1+n_2]$ for which the restrictions to the subsets

$$
\{1, \ldots, i-1\} \cup \{n_2+i, \ldots, n_1+n_2\} \text{ and } \{i, \ldots, i+n_2-1\}
$$

are order-preserving. The map τ interchanges the two middle tensor factors. In particular, if $i = n_1 + 1$, (8.16) is the summation over all (n_1, n_2)-unshuffles $\rho \in \Sigma_{n_1+n_2}$, i.e. permutations $\rho \in \Sigma_{n_1+n_2}$ such that

$$\rho(1) < \cdots < \rho(n_1) \text{ and } \rho(n_1 + 1) < \cdots < \rho(n_1 + n_2).$$

Proof. Consider the map $\theta : [n+2] \to [n]\sqcup\{i, j\}$ which coincides with τ_{ij} of (6.69) on $[n + 2] \setminus \{i, j\}$, while $\theta(i) := i$ and $\theta(j) := j$. By the functoriality (6.53) one has the equality

$$\circ_{ij}\mathcal{M}(\theta) = \mathcal{M}(\tau_{ij})\circ_{ij}$$

of maps $\mathcal{M}(n + 2; g) \to \mathcal{M}(n; g)$. On the other hand, $\mathcal{M}(\tau_{ij})\circ_{ij} = \bar{\circ}_{ij}$ by the definition (6.70) of skeletal operations, thus $\circ_{ij}\mathcal{M}(\theta) = \bar{\circ}_{ij}$. Likewise we establish that $\bullet_{ij}\mathcal{T}(\theta) = \bar{\bullet}_{ij}$. Formula (8.15) is then obtained by taking in (8.2) the above isomorphism θ.

Let us prove (8.16). Recall that the summation (8.3) defining the bracket runs over all subsets $S_1, S_2 \subset [n_1 + n_2]$ such that $S_1 \sqcup S_2 = [n_1 + n_2]$ and $|S_1| = n_1$, $|S_2| = n_2$. Let X_1, X_2 be two such subsets. There clearly exists a one-to-one correspondence between couples (S_1, S_2) and automorphisms $\rho : [n_1+n_2] \xrightarrow{\cong} [n_1+n_2]$ whose restrictions to X_1 and X_2 are order-preserving. It follows from this observation and the functoriality of the structure operations ${}_a\circ_b$ and ${}_a\bullet_b$ that, for fixed bijections

$$\vartheta_1 : [n_1 + 1] \xrightarrow{\cong} X_1 \sqcup \{a\} \text{ and } \vartheta_2 : [n_2 + 1] \xrightarrow{\cong} X_2 \sqcup \{b\} \tag{8.17}$$

the right-hand side of (8.3) equals the sum

$$\sum_{\rho} \big(\mathcal{M}(\rho) \otimes \mathcal{T}(\rho)\big)\big({}_a\circ_b \otimes {}_a\bullet_b\big)\big(\vartheta_1 \otimes \vartheta_2 \otimes \vartheta_1 \otimes \vartheta_2\big)\tau(g \otimes h) \tag{8.18}$$

over automorphisms $\rho : [n_1 + n_2] \xrightarrow{\cong} [n_1 + n_2]$ as above.

Let κ_{ij} be as in (6.7) with $m := n_1$ and $n := n_2$. Take

$$X_1 := \{1, \ldots, i - 1\} \cup \{n_2 + i, \ldots, n_1 + n_2\},\ X_2 := \{i, \ldots, i + n_2 - 1\}$$

and notice that

$$X_1 = \kappa_{ij}\big([n_1 + 1] \setminus \{i\}\big) \text{ and } X_2 = \kappa_{ij}\big([m_1 + 1] \setminus \{j\}\big).$$

Take as ϑ_1 in (8.17) the isomorphism that restricts to κ_{ij} on $[n_1 + 1] \setminus \{i\}$ and sends $i \in [n_1 + 1]$ to a. The isomorphism ϑ_2 is defined similarly. Recalling the definition (6.8) of the skeletal operations, we see that

$${}_a\circ_b(\vartheta_1 \otimes \vartheta_2) = \mathcal{M}(\kappa_{ij})\,{}_i\circ_j = {}_i\bar{\circ}_j$$

and, likewise,

$$_a\bullet_b(\vartheta_1 \otimes \vartheta_2) = \mathscr{T}(\kappa_{ij})\,{}_i\bullet_j = {}_i\overline{\bullet}_j\,.$$

Taking this in account, we see that for the choices of ϑ_1 and ϑ_2 above, (8.18) equals the sum in the right-hand side of (8.16) as required.

Suppose that $\mathscr{M}$ is a finite-dimensional, in the sense of Definition 7.8, modular operad with structure operations ${}_a\circ_b$ and $\circ_{uv}$. Suppose moreover that for each finite sets $S_1, S_2 \in \mathtt{Cor}$ and $g \in \mathbb{A}$, there are only finitely many couples $(g_1, g_2) \in \mathbb{A}^{\times 2}$ such that $g_1 + g_2 = g$ for which the restriction

$${}^{g_1}_{a}\circ^{g_2}_{b} : \mathscr{M}\big(S_1 \sqcup \{a\}; g_1\big) \otimes \mathscr{M}\big(S_2 \sqcup \{b\}; g_2\big) \to \mathscr{M}(S_1 \sqcup S_2; g_1 + g_2)$$

of the structure operation ${}_a\circ_b$ is non-zero. Such $\mathscr{M}$ clearly fulfills the assumptions of Proposition 7.1, therefore its piece-wise linear dual $\mathscr{M}^{\#}$ is a modular cooperad, with structure operations ${}_{S_1;g_1}\overset{a\;b}{\circ}{}_{S_2;g_2} := {}_a\circ_b{}^{\#}$ and $\circ^{uv}_{g} := \circ_{uv}{}^{\#}$.

The following statement rephrases Proposition 7.2 for the case when $\mathscr{C}$ is the piece-wise linear dual of a modular operad $\mathscr{M}$ as above.

Theorem 8.2 ([1]) *Assume that $\mathscr{M}$ is a finite-dimensional modular operad as above and $\mathscr{M}^{\#}$ its dual modular cooperad. A morphism*

$$\alpha : \mathscr{F}(\mathscr{M}^{\#}) \to \mathscr{T} \tag{8.19}$$

of odd modular dg-operads is then the same as a degree 0 element $S \in MT$ satisfying the master equation

$$d(S) + \Delta(S) + \frac{1}{2}\{S, S\} = 0 \tag{8.20}$$

in the desuspended bi-differential graded Lie algebra structure of Theorem 8.1.

Proof. The left-hand side of the master equation (8.20), as an element of the product MT, vanishes if and only if each of its factors does. The factor in arity n and genus g equals

$$dS(n; g) + \sum_{g'+s=g} \Delta S(n+2; g') + \frac{1}{2} \sum_{n_1+n_2=n}\ \sum_{g_1+g_2=g} \big\{S(n_1+1; g_1), S(n_2+1; g_2)\big\}. \tag{8.21}$$

We therefore need to prove that (8.21) vanishes for each $(n, g) \in \mathbb{N} \times \mathbb{A}$.

Let us start by recalling elementary facts about the linear duality. For graded vector spaces V and W, one has the canonical embedding

$$N : V \otimes W \hookrightarrow Lin(V^\#, W) \tag{8.22}$$

given, for homogeneous $\alpha \in V^\#$, $v \in V$ and $w \in W$, by

$$N(\alpha)(v \otimes w) = \alpha(v)w.$$

This embedding is functorial in the sense that, for $S \in V' \otimes W'$ and linear maps $h : V' \to V''$, $f : W' \to W''$,

$$N\big((h \otimes f)(S)\big) = f \circ N(S) \circ h^\# \in Lin(V'', W''). \tag{8.23}$$

If V is non-graded,[4] (8.22) is an isomorphism if and only if it is finite-dimensional.

In the general case, the situation is more complicated. The kth graded component of $V \otimes W$ equals

$$(V \otimes W)^k = \bigoplus_{i+j=k} V^i \otimes W^j,$$

while the part of $Lin(V^\#, W)$ of degree k equals

$$Lin(V^\#, W)^k = \prod_{i+j=k} Lin\big((V^\#)^{-i}, W^j\big) = \prod_{i+j=k} Lin\big((V^i)^\#, W^j\big). \tag{8.24}$$

We see that (8.22) need not be an isomorphism even when both V and W are of finite type. On the other hand, when both V and W are finite-dimensional, the product in (8.24) has only finite number of nontrivial factors, so it equals the direct sum, and (8.22) is an isomorphism.

If V and W are finite-dimensional graded left modules over a finite group G, then (8.22) restricts to an isomorphism

$$N = N_G : (V \otimes W)^G \to Lin_G(V^\#, W),$$

where $(V \otimes W)^G$ is the subspace of G-stable vectors under the diagonal action of G, and $Lin_G(V^\#, W)$ the subspace of linear maps $\phi : V^\# \to W$ which are equivariant in the sense that for each $\alpha \in V^\#$ and $g \in G$,

$$\phi(\alpha) = g\,\phi(\alpha g),$$

[4] That is, concentrated in degree 0.

where $\alpha \mapsto \alpha g$ is the action dual to the action of G on V. Notice also that for finite-dimensional vector spaces V_1, V_2, W_1, and W_2 one has the following diagram of natural isomorphisms:

$$\begin{array}{ccc} \otimes V_2 \otimes W_1 \otimes W_2 & \xrightarrow{N} & Lin\big((V_1 \otimes V_2)^{\#}, W_1 \otimes W_2\big) \\ {\scriptstyle \tau}\big\downarrow & & \big\uparrow{\scriptstyle \Upsilon} \\ \otimes W_1 \otimes V_2 \otimes W_2 & \xrightarrow{N \otimes N} & Lin(V_1^{\#}, W_1) \otimes Lin(V_2^{\#}, W_2). \end{array} \tag{8.25}$$

Let ${}_a\circ_b$ and $\circ_{uv}$ be the structure operations of the operad $\mathscr{M}$, and ${}_a\bullet_b$ and $\bullet_{uv}$ the structure operations of the odd modular operad $\mathscr{T}$. We start the actual proof by representing the morphism (8.19) by the skeletal version (7.36) of the family (7.31) with $\mathscr{C} = \mathscr{M}^{\#}$. By the finite-dimensionality assumption, for each $n \in \mathbb{N}$ and $g \in \mathbb{A}$, there exists a unique

$$S(n; g) \in MT(n; g) = \big(\mathscr{M}(n; g) \otimes \mathscr{T}(n; g)\big)^{\Sigma_n}$$

such that

$$N(S(n; g)) = A(n; g) : \mathscr{M}(n; g)^{\#} \to \mathscr{T}(n; g).$$

One therefore has a one-to-one correspondence between skeletal families $A_{\mathrm{sk}} = \{A(n; g) \mid (n, g) \in \mathbb{N} \times \mathbb{A}\}$ and degree 0 elements $S = \{S(n; g) \mid (n, g) \in \mathbb{N} \times \mathbb{A}\} \in MT$.

We need to prove that A_{sk} satisfies (7.39) if and only if the corresponding S satisfies the master equation (8.20). Let us denote, in (7.39) with $\mathscr{C} = \mathscr{M}^{\#}$,

$$U(n; g') := \bullet_{uv}\mathscr{T}(\theta)A(n+2; g')\mathscr{M}(\theta)^{\#}{\circ_{uv}}^{\#}, \quad \text{and}$$

$$V(n_1, n_2; g_1, g_2) := {}_a\bullet_b(\theta_1 \otimes \theta_2)\big(A(n_1{+}1; g_1) \otimes A(n_2{+}1; g_2)\big)(\theta_1^{\#} \otimes \theta_2^{\#})\,{{}_a\circ_b}^{\#}.$$

With this notation, Eq. (7.39) reads

$$\begin{aligned} d_{\mathscr{T}} A(n; g) = A(n; g)d^{\#}_{\mathscr{M}} &+ \sum_{g'+s=g} U(n; g') \\ &+ \frac{1}{2} \sum_{g_1+g_2=g} \; \sum_{S_1 \sqcup S_2 = [n]} V(n_1, n_2; g_1, g_2). \end{aligned} \tag{8.26}$$

Assume we have proved that

$$N\big(dS(n;g)\big) = A(n;g)d^{\#}_{\mathscr{M}} - d_{\mathscr{T}}A(n;g), \tag{8.27}$$

$$N\big(\Delta S(n+2;g')\big) = U(n;g), \quad \text{and} \tag{8.28}$$

$$N\big(\{S(n_1;g_1), S(n_2;g_2)\}\big) = V(n_1,n_2;g_1,g_2). \tag{8.29}$$

Then (8.26) is N applied to (8.21). Since N is an isomorphism by assumptions, (8.21) vanishes if and only if (8.26) does.

Let us prove (8.27). By the definition of the differential d,

$$N\big(dS(n;g)\big) = N\big((d_{\mathscr{M}} \otimes \mathbb{1})S(n;g)\big) - N\big((\mathbb{1} \otimes d_{\mathscr{T}})S(n;g)\big).$$

Taking (8.23) with $V' = V'' = \mathscr{M}(n;g)$, $W' = W'' = \mathscr{T}(n;g)$, $h = d_{\mathscr{M}}$, and $f = \mathbb{1}$ one sees that

$$N\big((d_{\mathscr{M}} \otimes \mathbb{1})S(n;g)\big) = N\big(S(n;g)\big)d^{\#}_{\mathscr{M}} = A(n;g)d^{\#}_{\mathscr{M}}$$

while, with $h = \mathbb{1}$ and $f = d_{\mathscr{T}}$, (8.23) gives

$$N\big((\mathbb{1} \otimes d_{\mathscr{T}})S(n;g)\big) = d_{\mathscr{T}}N\big(S(n;g)\big) = d_{\mathscr{T}}A(n;g).$$

The above three displays combine to (8.27). Equation (8.28) is proven by taking in (8.23)

$$V' = \mathscr{M}(n+2;g'),\ W' = \mathscr{T}(n+2;g'),\ V'' = \mathscr{M}(n;g'+s),$$
$$W'' = \mathscr{T}(n;g'+s),\ f = \bullet_{uv}\mathscr{T}(\theta) \text{ and } h = \circ^{\mathscr{M}}_{uv}\mathscr{M}(\theta).$$

By the definition (8.2) of the operation Δ, one has

$$\begin{aligned} N\big(\Delta S(n+2;g')\big) &= N\big((\circ^{\mathscr{M}}_{uv}\mathscr{M}(\theta) \otimes \bullet_{uv}\mathscr{T}(\theta))S(n+2;g')\big) \\ &= \bullet_{uv}\mathscr{T}(\theta)N(S(n+2;g'))\mathscr{M}(\theta)^{\#}{\circ_{uv}}^{\#} \\ &= \bullet_{uv}\mathscr{T}(\theta)A(n+2;g')\mathscr{M}(\theta)^{\#}{\circ_{uv}}^{\#} = U(n;g'), \end{aligned}$$

which is (8.28). Finally, by the definition (8.3) of the bracket,

$$\begin{aligned} &N\{S(n_1{+}1;g_1), S(n_2{+}1;g_2)\} \\ &= \sum_{S_1 \sqcup S_2 = [n_1+n_2]} N\big(({}_a\circ_b \otimes {}_a\bullet_b)\tau(\theta_1 \otimes \theta_1 \otimes \theta_2 \otimes \theta_2)(S(n_1{+}1;g_1) \otimes S(n_2{+}1;g_2))\big). \end{aligned} \tag{8.30}$$

As in the proof of Theorem 8.1, the above formula was shortened by denoting the actions of morphisms $\mathscr{M}(\theta_i)$ resp. $\mathscr{T}(\theta_i)$ by θ_i, $i = 1, 2$, the exact meaning being always clear from the context. For instance, the full name of the term in the bracket on which τ acts is

$$\mathscr{M}(\theta_1) \otimes \mathscr{T}(\theta_1) \otimes \mathscr{M}(\theta_2) \otimes \mathscr{T}(\theta_2).$$

Isomorphism (8.23) with

$$V' = \mathscr{M}(n_1+1; g_1) \otimes \mathscr{M}(n_2+1; g_2),\ W' = \mathscr{T}(n_1+1; g_1) \otimes \mathscr{T}(n_2+1; g_2)$$
$$V'' = \mathscr{M}(n; g),\ W'' = \mathscr{T}(n; g),\ h = {}_a\circ_b,\ \text{ and } f = {}_a\bullet_b,$$

converts the right-hand side of (8.30) into

$$\sum_{S_1 \sqcup S_2 = [n_1+n_2]} {}_a\bullet_b\, N\big(\tau(\theta_1 \otimes \theta_1 \otimes \theta_2 \otimes \theta_2)(S(n_1+1; g_1) \otimes S(n_2+1; g_2))\big)\, {}_a\circ_b{}^{\#}. \tag{8.31}$$

The commutativity of diagram (8.25), with

$$V_1 = \mathscr{M}(n_1+1; g_1),\ V_2 = \mathscr{M}(n_2+1; g_2),$$
$$W_1 = \mathscr{T}(n_1+1; g_1) \text{ and } W_2 = \mathscr{T}(n_2+1; g_2)$$

implies that

$$N\big(\tau(\theta_1 \otimes \theta_1 \otimes \theta_2 \otimes \theta_2)(S(n_1+1; g_1) \otimes S(n_2+1; g_2))\big)$$

equals

$$\Upsilon\big(N((\theta_1 \otimes \theta_1)(S(n_1+1; g_1))) \otimes N((\theta_1 \otimes \theta_1)(S(n_2+1; g_2)))\big)$$

which can be rewritten using (8.23) twice as

$$\Upsilon\big(\theta_1 A(n_1+1; g_1)\theta_1^{\#} \otimes \theta_2 A(n_2+1; g_1)\theta_2^{\#}\big)$$

which clearly equals

$$(\theta_1 \otimes \theta_2)\big(A(n_1+1; g_1) \otimes A(n_2+1; g_2)\big)(\theta_1^{\#} \otimes \theta_2^{\#}).$$

Inserting this expression into (8.31) gives the right-hand side of (8.29). This finishes the proof.

8.2 Loop Homotopy Algebras

In this section we describe algebras over the Feynman transform $\mathscr{F}(\mathscr{QC}^{\#})$ of the dual of the quantum closed operad $\mathscr{QC}$ and identify them with loop (aka quantum) homotopy Lie algebras. As we noticed in Examples 6.24 and 6.26, the modular operad $\mathscr{QC}$ is stable, isomorphic to the modular envelope $\mathrm{Mod}(\mathscr{C}om)$ of the cyclic commutative operad, which is in turn isomorphic to the linear span of the terminal stable modular operad $*_{\mathrm{Mod}}$. Thus

$$\mathscr{QC}(S; g) = \begin{cases} \Bbbk & \text{if } (S, g) \in \mathfrak{S}, \text{ and} \\ 0 & \text{otherwise,} \end{cases} \tag{8.32}$$

with the trivial actions of the symmetric groups. This implies that $\mathscr{QC}(S; g)^{\#} = \Bbbk$ for each $(S, g) \in \mathfrak{S}$ and that all the structure operations of the dual modular cooperad $\mathscr{QC}^{\#}$ are either the canonical isomorphisms $\Bbbk \xrightarrow{\cong} \Bbbk \otimes \Bbbk$ or the identities $\mathbb{1} : \Bbbk \xrightarrow{\cong} \Bbbk$.

As the first step, we analyze the space $MT(n; g)$ of Definition 8.1 in the case when $\mathscr{M} = \mathscr{QC}$ and when $\mathscr{T}$ is an odd modular operad with $\mathbb{A} = \mathbb{N}$ and step $s = 1$. It follows from (8.32) that

$$\mathrm{QCT}(n; g) = \begin{cases} \mathscr{T}(n; g)^{\Sigma_n} & \text{if } (n, g) \in \mathfrak{S}_{\mathrm{sk}}, \text{ and} \\ 0 & \text{otherwise,} \end{cases}$$

where $\mathfrak{S}_{\mathrm{sk}}$ is the skeletal version (6.71) of the set $\mathfrak{S}$, therefore

$$\mathrm{QCT} = \prod_{(n,g) \in \mathfrak{S}_{\mathrm{sk}}} \mathscr{T}(n; g)^{\Sigma_n}. \tag{8.33}$$

The skeletal form of the structure operations in Proposition 8.1 is easy to describe. Choosing $i = 1$, $j = 2$ in (6.12) gives

$$\Delta(f) = \bar{\bullet}_{12}(f) \in \mathscr{T}(n; g + 1)^{\Sigma_n} \tag{8.34}$$

for $f \in \mathscr{T}(n + 2; g)^{\Sigma_{n+2}}$. Similarly, (8.16) with $i = n_1 + 1$ and $j = 1$ gives[5]

$$\{g, h\} := \sum_{\sigma \in \mathrm{uSh}(n_1, n_2)} \mathscr{T}(\sigma)(g\, {}_{n_1+1}\bar{\bullet}_1\, h) \in \mathscr{T}(n_1 + n_2; g_1 + g_2)^{\Sigma_{n_1+n_2}} \tag{8.35}$$

[5] In Eq. (8.35), $g\, {}_{n_1+1}\bar{\bullet}_1\, h := {}_{n_1+1}\bar{\bullet}_1(g \otimes h)$.

for $g \in \mathcal{T}(n_1+1; g_1)^{\Sigma_{n_1+1}}$ and $h \in \mathcal{T}(n_2+2; g_2)^{\Sigma_{n_2+1}}$, with the summation running over all (n_1, n_2)-unshuffles $\sigma \in \Sigma_{n_1+n_2}$, that is, permutations σ such that

$$\sigma(1) < \cdots < \sigma(n_1) \text{ and } \sigma(n_1+1) < \cdots < \sigma(n_1+n_2).$$

The master equation (8.20) describing morphisms $\alpha : \mathcal{F}(\mathcal{Q}\mathcal{C}^{\#}) \to \mathcal{T}$ in terms of degree 0 elements $S \in \mathrm{QCT}$ therefore reads

$$d_{\mathcal{T}} S = \bar{\bullet}_{12}(S) + \frac{1}{2} \sum_{\sigma \in \mathrm{uSh}(n_1,n_2)} \mathcal{T}(\sigma)(S\,{}_{n_1+1}\bar{\bullet}_1\, S).$$

By (8.33), S is a sequence of elements $S(n; g) \in \mathcal{T}(n; g)^{\Sigma_n}$, $(n, g) \in \mathfrak{S}_{\mathrm{sk}}$, where the set $\mathfrak{S}_{\mathrm{sk}}$ was defined in (6.71). It will be convenient to put by definition $S(n; g) := 0$ if $(n, g) \notin \mathfrak{S}_{\mathrm{sk}}$. The above master equation then means that, for each $(n, g) \in \mathfrak{S}_{\mathrm{sk}}$,

$$d_{\mathcal{T}} S(n; g) = \bar{\bullet}_{12} S(n+2; g-1) + \frac{1}{2} \sum S(n_1+1; g_1)\,{}_{n_1+1}\bar{\bullet}_1\, S(n_2+1; g_2) \tag{8.36}$$

with the summation taken over $n_1+n_2 = n$, $g_1+g_2 = g$ and $\sigma \in \mathrm{uSh}(n_1, n_2)$.

Let us finally apply our machinery in the situation when $\mathcal{T}$ is the odd endomorphism operad $\mathcal{E}nd_V$ of Example 6.31 in the skeletal presentation given in Example 6.32. In this particular case we use more traditional notation and denote $f_n^g := S(n; g) \in \mathcal{E}nd_V(n; g)$, for $n \geq 0$, $g \geq 0$. So f_n^g is a degree 0 function $V^{\otimes n} \to \Bbbk$ which is zero if $(n, g) \notin \mathfrak{S}_{\mathrm{sk}}$. If V is equipped with a differential d, we denote by the same symbol the differential induced in the standard manner on the space of functions $V^{\otimes n} \to \Bbbk$.

Theorem 8.3 *An algebra over the Feynman transform $\mathcal{F}(\mathcal{Q}\mathcal{C}^{\#})$ on a dg-vector space $V = (V, d)$ equipped with a degree $+1$ d-closed symmetric element*

$$s = \sum s_i' \otimes s_i'' \in V \otimes V$$

is the same as a collection

$$\left\{ f_n^g : V^{\otimes n} \to \Bbbk \mid (n, g) \in \mathfrak{S}_{\mathrm{sk}} \right\}$$

of degree 0 totally symmetric linear maps satisfying, for each $(n, g) \in \mathfrak{S}_{\mathrm{sk}}$, the equation

$$d(f_n^g) = \sum f_{n+2}^{g-1}(s_i' \otimes s_i'' \otimes \mathbb{1}^{\otimes n}) \tag{8.37}$$
$$+ \frac{1}{2} \sum \left(f_{n_1+1}^{g_1}(\mathbb{1}^{\otimes n_1} \otimes s_i') f_{n_2+1}^{g_2}(s_i'' \otimes \mathbb{1}^{\otimes n_2}) \right) \sigma^{-1},$$

where the second sum is taken over all $n_1 + n_2 = n$, $g_1 + g_2 = g$, *and unshuffles* $\sigma \in \mathrm{uSh}(n_1, n_2)$.[6]

Proof. All terms in (8.37) are degree $+1$ totally symmetric functions $V^{\otimes n} \to \Bbbk$. Let us inspect how they act on a homogeneous element $v_1 \otimes \cdots \otimes v_n \in V^{\otimes n}$. As before, we will save space by writing e.g. $f_n^g(v_1, \ldots, v_n)$ instead of $f_n^g(v_1 \otimes \cdots \otimes v_n)$. The value of the first term in the right-hand side of (8.37) is

$$\sum f_{n+2}^{g-1}(s_i' \otimes s_i'' \otimes \mathbb{1}^{\otimes n})(v_1, \ldots, v_n) = \sum f_{n+2}^{g-1}(s_i', s_i'', v_1, \ldots, v_n)$$

which coincides with $\bar{\bullet}_{12}(f_{n+2}^{g-1})$ described in Example 6.32, evaluated at $v_1, \ldots, v_n$.

The symbol σ^{-1} occurring in the second term in the right-hand side denotes the map $V^{\otimes n} \to V^{\otimes n}$ that permutes the factors of $V^{\otimes n}$ according to the permutation σ^{-1}. Thus

$$\frac{1}{2}\sum \big(f_{n_1+1}^{g_1}(\mathbb{1}^{\otimes n_1} \otimes s_i') f_{n_2+1}^{g_2}(s_i'' \otimes \mathbb{1}^{\otimes n_2})\big)\sigma^{-1}(v_1, \ldots, v_n)$$
$$= \frac{1}{2}\sum \epsilon(\sigma)(-1)^{|s_i'|} f_{n_1+1}^{g_1}(v_{\sigma(1)}, \ldots, v_{\sigma(n_1)}, s_i') f_{n_2+1}^{g_2}(s_i'', v_{\sigma(n_1+1)}, \ldots, v_{\sigma(n)}),$$

where $\epsilon(\sigma)$ is the Koszul sign of the permutation σ. The summand obviously, modulo the sign factor, coincides with

$$\mathscr{E}nd_V(\sigma)(f_{n_1+1}^{g_1}\ {}_{n_1+1}\bar{\bullet}_1\ f_{n_2+1}^{g_2})$$

as defined in Example 6.32, evaluated at $v_1, \ldots, v_n$. To verify that also the sign factor $(-1)^{|s_i'|}$ is in place, we need to realize that, since $f_{n_1+1}^{g_1}$ is of degree 0,

$$f_{n_1+1}^{g_1}(v_{\sigma(1)}, \ldots, v_{\sigma(n_1)}, s_i') \neq 0$$

only if

$$|v_{\sigma(1)}| + \cdots + |v_{\sigma(n_1)}| + |s_i'| = 0,$$

therefore $|v_{\sigma(1)}| + \cdots + |v_{\sigma(n_1)}| \equiv |s_i'| \mod 2$. We thus verified that (8.37) is indeed (8.36) with $\mathscr{T} = \mathscr{E}nd_V$.

For the sake of completeness we add that the term in left-hand side of (8.37) acts, by the definition of the induced differential, on $v_1 \otimes \cdots \otimes v_n$ by

$$d(f_n^g)(v_1, \ldots, v_n) = -\sum_{1 \leq i \leq n} (-1)^{|v_1| + \cdots + |v_{i-1}|} f_n^g(v_1, \ldots, dv_i, \ldots, v_n), \tag{8.38}$$

[6] The summation over repeated indexes, i.e. i in this case, is implicitly assumed.

We are going to rewrite (8.37) into an equivalent form closer to Equation (33) of [9]. Our translation will be based on the correspondence that assigns, for $n \geq 1$, to a degree k function $f : V^{\otimes n} \to \Bbbk$ the degree $k+1$ map $\delta : V^{\otimes(n-1)} \to V$ by the formula

$$\delta(v_1, \ldots, v_{n-1}) =: \sum (-1)^{k|s_i''|} f(s_i', v_1, \ldots, v_{n-1}) s_i'', \tag{8.39}$$

for $v_1, \ldots, v_{n-1} \in V$. The correspondence $f \mapsto \delta$ is one-to-one when s is non-degenerate. Let us explain the sign factor $(-1)^{k|s_i''|}$. We started from the composition[7]

$$\sum (\mathbb{1} \otimes f)(s_i'' \otimes s_i' \otimes v_1, \ldots, v_{n-1})$$

which, according to the Koszul sign convention, equals

$$\sum (-1)^{k|s_i''|} s_i'' \otimes f(s_i', v_1, \ldots, v_{n-1}).$$

Next we commuted the degree $|s_i''|$ vector s_i'' with the scalar $f(s_i', v_1, \ldots, v_{n-1})$ of degree 0 and obtained

$$\sum (-1)^{k|s_i''|} f(s_i', v_1, \ldots, v_{n-1}) \otimes s_i''.$$

Finally, we multiplied the vector s_i'' with the scalar $f(s_i', v_1, \ldots, v_{n-1})$. The result was the right-hand side of (8.39).

Notice that when f is fully symmetric in homogeneous $v_1, \ldots, v_n$, δ is fully symmetric in $v_1, \ldots, v_{n-1}$. In this case also

$$f(s_i', v_1, \ldots, v_{n-1}) = (-1)^{|s_i'|(|v_1| + \cdots + |v_{n-1}|)} f(v_1, \ldots, v_{n-1}, s_i').$$

Since f is of degree k, $f(v_1, \ldots, v_{n-1}, s_i') \neq 0$ only if

$$|v_1| + \cdots + |v_{n-1}| + |s_i'| + k = 0,$$

so $|v_1| + \cdots + |v_{n-1}| \equiv |s_i'| + k \mod 2$. We therefore have

$$\begin{aligned} f(s_i', v_1, \ldots, v_{n-1}) &= (-1)^{|s_i'|(|s_i'|+k)} f(v_1, \ldots, v_{n-1}, s_i') \\ &= (-1)^{|s_i'| + k|s_i'|} f(v_1, \ldots, v_{n-1}, s_i'). \end{aligned}$$

[7]Notice that $\sum s_i'' \otimes s_i' = s_i' \otimes s_i''$ because s is symmetric and $|s_i'| + |s_i''| = 1$.

Substituting the last expression into formula (8.39) produces, in the symmetric case, an alternative formula

$$\begin{aligned}\delta(v_1, \dots, v_{n-1}) &= (-1)^{|s_i'|+k(|s_i'|+|s_i''|)} \sum f(v_1, \dots, v_{n-1}, s_i')s_i'' \\ &= (-1)^{|s_i'|+k} \sum f(v_1, \dots, v_{n-1}, s_i')s_i''.\end{aligned} \tag{8.40}$$

Let us finally denote, for $n \geq 1$ and $g \geq 0$, by $\delta^g_{n-1} : V^{\otimes n-1} \to V$ the degree 1 map corresponding to the degree 0 function $f^g_n : V^n \to \Bbbk$. Explicitly

$$\begin{aligned}\delta^g_{n-1}(v_1, \dots, v_{n-1}) &:= \sum f^n_g(s_i', v_1, \dots, v_{n-1})s_i'' \\ &= (-1)^{|s_i'|} \sum f^n_g(v_1, \dots, v_{n-1}, s_i')s_i''\end{aligned}$$

for $v_1, \dots, v_{n-1} \in V$.

Let us start to apply our correspondence between functions $V^n \to \Bbbk$ and maps $V^{\otimes n-1} \to V$ to the terms of (8.37). Since they are all of degree 1, we have $k = 1$ in (8.39) resp. in (8.40). Using formula (8.38), we conclude that the map corresponding to the left-hand side of (8.37) is given by

$$\begin{aligned}&\sum_{1 \leq i \leq n-1} (-1)^{|v_1|+\dots+|v_{i-1}|+|s_i'|} f^g_n(v_1, \dots, dv_i, \dots, v_{n-1}, s_i')s_i'' \\ &- \sum f^g_n(v_1, , \dots, v_{n-1}, ds_i')s_i''.\end{aligned}$$

The rightmost term appears with the minus sign because $f^g_n(v_1, , \dots, v_{n-1}, ds_i') \neq 0$ only if

$$|v_1| + \dots + |v_{n-1}| + |s_i'| + 1 = 0.$$

Since s is d-closed by assumption, $ds_i' \otimes s_i'' + (-1)^{|s_i'|} s_i' \otimes ds_i'' = 0$, so

$$-f^g_n(v_1, \dots, v_{n-1}, ds_i')s_i'' = (-1)^{|s_i'|} f^g_n(v_1, \dots, v_{n-1}, s_i')ds_i''.$$

By (8.39),

$$\begin{aligned}\sum f^g_n(v_1, \dots, dv_i, \dots, v_{n-1}, s_i')s_i'' &= (-1)^{|s_i'|} \sum \delta^g_{n-1}(v_1, \dots, dv_i, \dots, v_{n-1}) \\ \text{and} \sum f^g_n(v_1, \dots, v_{n-1}, s_i')s_i'' &= (-1)^{|s_i'|} \sum \delta^g_{n-1}(v_1, \dots, v_{n-1})\end{aligned}$$

which, combined with the above calculations, shows that the map corresponding to the left-hand side of (8.37) is given by

$$\sum_{1\leq i\leq n-1} (-1)^{|v_1|+\cdots+|v_{i-1}|}\delta^g_{n-1}(v_1,\ldots,dv_i,\ldots,v_{n-1}) + d\delta^g_{n-1}(v_1,\ldots,v_{n-1}).$$

We conclude that the left-hand side of (8.37) is translated into $d(\delta^g_{n-1})$, where d now denotes the induced differential on the space of maps $V^{\otimes n-1} \to V$.

The degree 2 map $V^{\otimes(n-1)} \to V$ corresponding to the first term in the right-hand side of (8.37) is, by (8.39), given as

$$\begin{aligned}
&\sum (-1)^{|s''_j|} f^{g-1}_{n+2}(s'_i, s''_i, s'_j, v_1,\ldots,v_{n-1})s''_j \\
&= \sum (-1)^{|s''_j|+(|s'_i|+|s''_i|)|s'_j|} f^{g-1}_{n+2}(s'_j, s'_i, s''_i, v_1,\ldots,v_{n-1})s''_j \\
&= -\sum \delta^{g-1}_{n+1}(s'_i, s''_i, v_1,\ldots,v_{n-1}),
\end{aligned}$$

where we used that $|s''_j| + (|s'_i| + |s''_i|)|s'_j| = |s''_j| + |s'_j| = 1$.

The analysis of the second term in the right-hand side is subtler. Denote by $\mathrm{uSh}'(n_1,n_2)$ the set of all (n_1,n_2)-unshuffles $\sigma \in \mathrm{uSh}(n_1,n_2)$ such that $\sigma(n) = n$. Notice that then

$$\begin{aligned}
&\frac{1}{2}\sum \epsilon(\sigma)(-1)^{|s'_i|} f^{g_1}_{n_1+1}(v_{\sigma(1)},\ldots,v_{\sigma(n_1)},s'_i) f^{g_2}_{n_2+1}(s''_i, v_{\sigma(n_1+1)},\ldots,v_{\sigma(n)}) \\
&= \sum{}' \epsilon(\sigma)(-1)^{|s'_i|} f^{g_1}_{n_1+1}(v_{\sigma(1)},\ldots,v_{\sigma(n_1)},s'_i) \\
&\times f^{g_2}_{n_2+1}(s''_i, v_{\sigma(n_1+1)},\ldots,v_{\sigma(n-1)},v_n),
\end{aligned}$$

where $\sum'$ denotes the sum in the first line of the display restricted to $\mathrm{uSh}'(n_1,n_2)$. Consequently, the degree $+2$ map $V^{\otimes(n-1)} \to V$ corresponding to the second term in the right-hand side of (8.37) is

$$\begin{aligned}
&-\sum{}' \epsilon(\sigma)(-1)^{|s'_i|+|s'_j|} f^{g_1}_{n_1+1}(v_{\sigma(1)},\ldots,v_{\sigma(n_1)},s'_i) \\
&\times f^{g_2}_{n_2+1}(s''_i, v_{\sigma(n_1+1)},\ldots,v_{\sigma(n-1)},s'_j)s''_j.
\end{aligned}$$

Noticing that

$$f^{g_2}_{n_2+1}(s''_i, v_{\sigma(n_1+1)},\ldots,v_{\sigma(n-1)},s'_j)s''_j = (-1)^{|s'_j|}\delta^{g_2}_{n_2}(s''_i, v_{\sigma(n_1+1)},\ldots,v_{\sigma(n-1)}),$$

we rewrite it into

$$-\sum{}' \epsilon(\sigma)(-1)^{|s'_i|} f^{g_1}_{n_1+1}(v_{\sigma(1)},\ldots,v_{\sigma(n_1)},s'_i)\delta^{g_2}_{n_2}(s''_i, v_{\sigma(n_1+1)},\ldots,v_{\sigma(n-1)}).$$

Since $f^{g_1}_{n_1+1}(v_{\sigma(1)}, \ldots, v_{\sigma(n_1)}, s'_i) \in \Bbbk$, we use the multilinearity of $\delta^{g_2}_{n_2}$ and further rewrite the above sum into

$$-\sum{}' \epsilon(\sigma)(-1)^{|s'_i|}\delta^{g_2}_{n_2}\big(f^{g_1}_{n_1+1}(v_{\sigma(1)}, \ldots, v_{\sigma(n_1)}, s'_i)s''_i, v_{\sigma(n_1+1)}, \ldots, v_{\sigma(n-1)}\big).$$

Noticing finally that

$$f^{g_1}_{n_1+1}(v_{\sigma(1)}, \ldots, v_{\sigma(n_1)}, s'_i)s''_i = (-1)^{|s'_i|}\delta^{g_1}_{n_1}(v_{\sigma(1)}, \ldots, v_{\sigma(n_1)}),$$

we see that the above sum equals

$$-\sum{}' \epsilon(\sigma)\delta^{g_2}_{n_2}\big(\delta^{g_1}_{n_1}(v_{\sigma(1)}, \ldots, v_{\sigma(n_1)}), v_{\sigma(n_1+1)}, \ldots, v_{\sigma(n-1)}\big)$$

which in turn equals

$$-\sum \epsilon(\tau)\delta^{g_2}_{n_2}\big(\delta^{g_1}_{n_1}(v_{\tau(1)}, \ldots, v_{\tau(n_1)}), v_{\tau(n_1+1)}, \ldots, v_{\tau(n-1)}\big),$$

where τ runs over all $(n_1, n_2 - 1)$-unshuffles.

Summing the above calculations, we obtain the formula

$$\begin{aligned} 0 = {} & d(\delta^g_{n-1})(v_1, \ldots, v_{n-1}) + \sum \delta^{g-1}_{n+1}(s'_i, s''_i, v_1, \ldots, v_{n-1}) \\ & + \sum \epsilon(\tau)\delta^{g_2}_{n_2}\big(\delta^{g_1}_{n_1}(v_{\tau(1)}, \ldots, v_{\tau(n_1)}), v_{\tau(n_1+1)}, \ldots, v_{\tau(n-1)}\big). \end{aligned} \tag{8.41}$$

that has to be satisfied for all homogeneous $v_1, \ldots, v_{n-1} \in V$. Recall that the second summation in the right-hand side runs over all $n_1 + n_2 = n$, $g_1 + g_2 = g$, and unshuffles $\tau \in \mathrm{uSh}(n_1, n_2 - 1)$. In a concise, elements-free form (8.41) reads

$$0 = d(\delta^g_{n-1}) + \sum \delta^{g-1}_{n+1}(s'_i, s''_i, \mathbb{1}^{\otimes(n-1)}) + \sum \delta^{g_2}_{n_2}(\delta^{g_1}_{n_1}, \mathbb{1}^{\otimes(n_1-1)})\tau^{-1}.$$

By the stability assumption, $f^0_2 : V^{\otimes 2} \to V$ and therefore $\delta^0_1 : V \to V$ is identically zero. A useful trick is to define $\delta^0_1 := d$ to be the differential of the underlying vector space. The first term in the right-hand side of (8.41) can then be absorbed into the third one, leading to

$$\begin{aligned} 0 = {} & \sum \epsilon(\tau)\delta^{g_2}_{n_2}\big(\delta^{g_1}_{n_1}(v_{\tau(1)}, \ldots, v_{\tau(n_1)}), v_{\tau(n_1+1)}, \ldots, v_{\tau(n-1)}\big) \\ & + \sum \delta^{g-1}_{n+1}(s'_i, s''_i, v_1, \ldots, v_{n-1}). \end{aligned} \tag{8.42}$$

Changing the symbol n to $n+1$, n_1 to l, n_2 to k, g_1 to g_2, g_2 to g_1 and τ to σ, and replacing $\sum s'_i \otimes s''_i$ to $\frac{1}{2}\sum y_i \otimes y^i$ we recognize the incarnation of *loop homotopy algebras* described in [9, Sublemma 2]. Particular cases of this structure with $\delta^g_n = 0$

whenever $g \geq 1$ are *strongly homotopy Lie algebras* [8], see the discussion of the "tree level" in [9, page 372].

8.3 IBL_∞-algebras

In this final section of the mathematical part we review, following closely [12], a common generalization of loop homotopy algebras as well as Lie bialgebras. We start by recalling some necessary auxiliary notions. Let A be a unital associative commutative algebra and $\Delta : A \to A$ a $\Bbbk$-linear map. For $n \geq 0$, consider the iterated graded commutators

$$[[\ldots[\Delta, L_{a_1}], \ldots], L_{a_n}] : A \to A,$$

with L_a denoting the operator of left multiplication by $a \in A$. By convention, we just set the commutator of Δ with $n = 0$ left-multiplication operators to be Δ. We call an operator Δ an *order* $\leq k$ *differential operator* if the iterated commutator with any $k + 1$ left-multiplication operators vanish.

Now suppose that $\Delta(1) = 0$.[8] Define

$$\Phi_n^\Delta(a_1, \ldots, a_n) := [[\ldots[\Delta, L_{a_1}], \ldots], L_{a_n}](1) \in A.$$

In particular, $\Phi_0^\Delta = 0$. If $\Phi_n^\Delta = 0$ for $n > k$, the operator Δ is called an *order* k *derivation* [11, Section 1.2].

Example 8.1 Assume for simplicity that the degree of Δ is 0 and that A is ungraded, the general graded case can be discussed analogously. For $a, x \in A$ one has $[\Delta, L_a](x) = \Delta(ax) - a\Delta(x)$. Invoking the unitality of A we immediately see that Δ is an order 0 operator if and only if it is the left multiplication by $\Delta(1)$. Since $\Phi_1^\Delta = \Delta$, the only degree 0 derivations are trivial maps.

For $a, b, x \in A$ one has $\big[[\Delta, L_a], L_b\big](x) = \Delta(abx) - a\Delta(bx) - \Delta(ax)b + a\Delta(x)b$. We leave as an exercise to show that Δ is an order ≤ 1 operator if and only if it (uniquely) decomposes into the sum of the left multiplication $L_{\Delta(1)}$ with a derivation. Since

$$\Phi_2^\Delta(a, b) = \Delta(ab) - a\Delta(b) - \Delta(a)b,$$

the first order derivations are "ordinary" derivations, i.e. vector fields.

Definition 8.2 Let $S(U)$ be the polynomial algebra generated by a graded vector space U and $\hbar$ a formal degree 0 symbol. An *IBL_∞-algebra* structure on U [4], [13, §4.2] is given by a degree 1, $\Bbbk[[\hbar]]$-linear map $\Delta : S(U)[[\hbar]] \to S(U)[[\hbar]]$

[8] All operators Δ in this section will share this property.

satisfying $\Delta^2 = 0$, $\Delta(1) = 0$, which moreover decomposes into a sum

$$\Delta = \Delta_1 + \hbar\Delta_2 + \hbar^2\Delta_3 + \cdots, \tag{8.43}$$

where

$$\Delta_k : S(U)[[\hbar]] \to S(U)[[\hbar]] \tag{8.44}$$

is an order $\leq k$ differential operator on the polynomial algebra $S(U)[[\hbar]]$.

As proved in [9, Proposition 3], an order $\leq k$ differential operator as in (8.44) with $\Delta(1) = 0$ is determined by its restrictions

$$\Delta_k|_{S^t(U)[[\hbar]]} : S^t(U)[[\hbar]] \to S(U)[[\hbar]]$$

to the subspaces $S^t(U) \subset S(U)$ of polynomials of length t with $1 \leq t \leq k$. These restrictions are, due to the assumed $\Bbbk[[\hbar]]$-linearity, in turn determined by their restrictions

$$\omega_{k,t} := \Delta_k|_{S^t(U)} : S^t(U) \to S(U)[[\hbar]] \tag{8.45}$$

and thus, after singling out the coefficients at $\hbar^g$, by the family

$$\omega_{k,t,g} := \Delta_k|_{S^t(U)} : S^t(U) \to S(U)$$

such that $\omega_{k,t} = \sum_{g\geq 0} \omega_{k,t,g}\hbar^g$. Moreover, each $\omega_{k,t,g}$ is determined by a family

$$\omega^s_{k,t,g} : S^t(U) \to S^s(U), \ s \geq 0, \tag{8.46}$$

which is locally finite in the sense that, for a given $u \in S^t(U)$, $\omega^s_{k,t,g}(u) \neq 0$ for only finitely many s.

On the other hand, backtracking the above procedure we readily see that each family of locally finite degree $+1$ maps

$$\left\{\omega^s_{k,t,g} : S^t(U) \to S^s(U) \mid k \geq 1, \ 1 \leq t \leq k, \ g, s \geq 0\right\} \tag{8.47}$$

as in (8.46) assembles into a map Δ in (8.43). The one-to-one correspondence $\{\omega^s_{k,t,g}\} \leftrightarrow \Delta$ can be made explicit using the calculations in Section 3 of [9]. General formulas are however clumsy so we do not include them here.

Remark 8.1 We saw that IBL_∞-algebras can be viewed as structures with infinitely many degree $+1$ operations (8.46) satisfying an infinite set of axioms obtained by assembling them into Δ and requiring $\Delta^2 = 0$. Since IBL_∞-algebras possess structure operations with several inputs and several outputs, they are not algebras over operads, but over more general objects called PROPs recalled e.g. in [10].

One sometimes considers versions of Definition 8.2 with $\hbar$ having degree different from 0. The convention when $|\hbar| = 2$ is implicit in [6] and explicit in [2], an arbitrary even degree is allowed in [3, 4]. Our convention that $\hbar$ has degree zero follows [13] and, of course, [12]. When $|\hbar| \neq 0$, the operator Δ_k in (8.43) has degree

$$1 + (1 - k)|\hbar|$$

and the operators $\omega^s_{k,t,g}$ in (8.47) degree

$$1 + (1 - k - g)|\hbar|.$$

Example 8.2 Suppose that $|\hbar| = 2$ and consider IBL_∞-algebras for which the only nontrivial operations are $\omega^1_{2,2,0} : S^2(U) \to U$ of degree -1 and $\omega^2_{1,1,0} : U \to S^2(U)$ of degree 1. This in particular means that Δ in (8.43) is of the form

$$\Delta = \Delta_1 + \hbar\Delta_2, \tag{8.48}$$

where $\Delta_1, \Delta_2 : S(U) \to S(U)$. It turns out that such IBL_∞-algebras are the same as *involutive Lie bialgebras* whose definition we recall below, on the desuspension $V := \downarrow U$ of U.

A *Lie bialgebra* is a graded vector space V equipped with a Lie algebra structure $\ell = [-, -] : V \otimes V \to V$ and a Lie diagonal (comultiplication) $\delta : V \to V \otimes V$. Explicitly, we assume that the bracket $[-, -]$ is anti-symmetric and satisfies the Jacobi equation

$$(-1)^{|a||c|}\big[[a, b], c\big] + (-1)^{|c||b|}\big[[c, a], b\big] + (-1)^{|b||a|}\big[[b, c], a\big] = 0 \tag{8.49}$$

and that δ satisfies the obvious duals of these conditions. We also assume that $[-, -]$ and δ are related by

$$\begin{aligned}\delta[a, b] \\ = \sum \Big((-1)^{|a''_i||b|}[a'_i, b] \otimes a''_i + [a, b'_j] \otimes b''_j + a'_i \otimes [a''_i, b] \\ +(-1)^{|a||b'_j|} b'_j \otimes [a, b''_j]\Big)\end{aligned} \tag{8.50}$$

for any $a, b \in V$, where we used the Sweedler notation

$$\delta a = \sum a'_i \otimes a''_i, \ \text{ and } \delta b = \sum b'_j \otimes b''_j.$$

A Lie bialgebra as above is *involutive* if, moreover

$$\sum[a_i', a_i''] = 0, \tag{8.51}$$

for $a \in V$ and $\sum a_i' \otimes a_i'' := \delta a$.

Returning to an IBL_∞-algebra in the beginning of this example, one may define degree 0 linear operations on the desuspension $V = \downarrow U$ by

$$\ell := \downarrow \omega^1_{2,2,0}(\uparrow \otimes \Uparrow) : V \wedge V \to V \text{ and } \delta := (\downarrow \otimes \Downarrow)\, \omega^2_{1,1,0} \uparrow : V \to V \wedge V.$$

Due to the special form (8.48) of Δ, one has

$$\Delta^2 = (\Delta_1 + \hbar \Delta_2)^2 = \Delta_1^2 + (\Delta_1\Delta_2 + \Delta_2\Delta_1)\hbar + \Delta_2^2\hbar^2.$$

Therefore $\Delta^2 = 0$ is equivalent to the separate vanishing of Δ_1^2, $(\Delta_1\Delta_2 + \Delta_2\Delta_1)^2$, and Δ_2^2. Since Δ_1^2 is a derivation, see [9, Proposition 1], its vanishing is equivalent to the vanishing of the restriction $\Delta_1^2|_U : U \to S^3(U)$ which is easily seen to be equivalent to the dual Jacobi identity for δ. A similar reasoning shows that the vanishing of $(\Delta_1\Delta_2 + \Delta_2\Delta_1)^2$ is equivalent to the compatibility (8.50) and the involutivity (8.51). Finally, the vanishing of Δ_2^2 is equivalent to the Jacobi identity for $[-,-] = \ell$. With this example in mind, one might view IBL_∞-algebras as homotopy versions of involutive Lie bialgebras, which explains the terminology.

Example 8.3 In this and the following example we assume that U is a graded vector space with finite-dimensional components. With this assumption, IBL_∞-algebras whose only nontrivial operations are

$$\omega^s_{1,1,0} : U \to S^s(U),\ s \geq 1,$$

are the same as L_∞-algebras on the dual $V = U^\#$ of U. Indeed, the operator Δ in (8.43) is in this case just a derivation $\Delta_1 : S(U) \to S(U)$ such that $\Delta_1^2 = 0$. Our statement is then the second part of [8, Theorem 2.3] with the reversed grading. Notice that $d := (\omega^1_{1,1,0})^\# : V \to V$ is a degree $+1$ differential.

Example 8.4 We leave as an exercise to prove that IBL_∞-algebras whose only nontrivial operations are

$$\omega^1_{1,1,0} : U \to U,\ \left\{\omega^s_{1,1,g} : U \to S^s(U)\right\}_{s\geq 2, g\geq 0} \text{ and } B := \omega^0_{2,2,0} : S^2(U) \to \Bbbk \tag{8.52}$$

are the same as loop homotopy Lie algebras on the dual $V := U^\#$ in the form (8.42). Indeed, defining the differential d on V to be the dual of $\omega^1_{1,1,0}$, the map $s : \Bbbk \to S^2(V)$ the dual of B and the operations δ^g_n the duals of $\omega^s_{1,1,g}$, we see that the structure equation $\Delta^2 = 0$ for Δ assembling the operations in (8.52) is equivalent to the axioms of a loop homotopy algebra with the operations δ_n^g acting on the differential graded vector space (V, d) equipped with the closed symmetric degree $+1$ element $s \in V \otimes V$.

The basic difference between the description of loop homotopy algebras given in Sect. 8.2 and the one in Example 8.4 above is that there the bilinear form $B = \omega_{2,2,0}^0$ is considered as a *structure operation*, while in Sect. 8.2 the corresponding symmetric element was fixed from the beginning.

Morphism Besides the commutative associative multiplication, the polynomial algebra $S(U)$ bears also the *coproduct* $\delta : S(U) \to S(U) \otimes S(U)$ which turns it into a commutative associative cocommutative coassociative coalgebra. The coproduct given by the formula

$$\delta(u_1 \odot \cdots \odot u_n) = \sum \frac{\epsilon(\sigma)}{a!\, b!}[u_{\sigma(1)} \odot \cdots \odot u_{\sigma(a)}] \otimes \cdots \otimes [u_{\sigma(a+1)} \odot \cdots \odot u_{\sigma(n)}],$$

where the summation runs over all permutations $\sigma \in \Sigma_k$ and integers $a, b \geq 0$ such that $a + b = n$. As always, $\epsilon(\sigma)$ denotes the Koszul sign and by $\odot$ we denote the standard commutative product of $S(U)$.

It is well known, see e.g. [5, §III.3], that the space $Lin_{\mathbb{k}}(A, C)$ of linear maps from a commutative associative algebra A with multiplication μ to a cocommutative coassociative coalgebra C with comultiplication δ admits the commutative coassociative *convolution product* $\star$ defined as

$$f \star g := \mu(f \otimes g)\delta, \;\; f, g \in Lin_{\mathbb{k}}(A, C).$$

In particular, $Lin_{\mathbb{k}}\big(S(U'), S(U'')\big)$ with the convolution product $\star$ is a commutative associative algebra. One easily sees that the composition

$$\mathsf{e} := S(U') \twoheadrightarrow S^0(U') \cong \mathbb{k} \cong S^0(U'') \hookrightarrow S(U'') \in Lin_{\mathbb{k}}\big(S(U'), S(U'')\big)$$

of the natural projection followed by the natural inclusion is the unit for $\star$. The above constructions extend by the $\mathbb{k}[[\hbar]]$-linearity to the space

$$Lin_{\mathbb{k}[[\hbar]]}\big(S(U')[[\hbar]], S(U'')[[\hbar]]\big)$$

of $\mathbb{k}[[\hbar]]$-linear maps which we will, for brevity, denote by $Lin_\hbar\big(S(U'), S(U'')\big)$ believing that the reader will not be too confused by this shorthand.

We denote finally by $Lin_\hbar^0\big(S(U'), S(U'')\big)$ the subset of $Lin_\hbar\big(S(U'), S(U'')\big)$ consisting of $\mathbb{k}[[\hbar]]$-linear maps such that

$$f(1) \in S(U'')[[\hbar]]. \tag{8.53}$$

The conilpotency of the coalgebra structure of $S(U')$ together with the $\hbar$-adic completeness of $\Bbbk[[\hbar]]$ implies:

Lemma 8.2 *All power series in elements of* $Lin^0_\hbar(S(U'), S(U''))$ *converge.*[9]

In particular, for $f \in Lin^0_\hbar(S(U'), S(U''))$ it makes sense to take the exponential

$$\exp(f) := \mathsf{e} + f + \frac{f^2}{2!} + \frac{f^3}{3!} + \cdots \in Lin^0_\hbar(S(U'), S(U''))$$

as well as the logarithm

$$\log(\mathsf{e} + f) := f - \frac{f^2}{2} + \frac{f^3}{3} - \cdots \in Lin^0_\hbar(S(U'), S(U'')).$$

Having prepared this auxiliary material, we formulate:

Definition 8.3 A *morphism* of IBL_∞-algebras $(S(U'), \Delta')$ and $(S(U''), \Delta'')$ is a $\Bbbk[[\hbar]]$-linear map

$$f : S(U')[[\hbar]] \to S(U'')[[\hbar]]$$

of the form

$$f = f^{(1)} + \hbar f^{(2)} + \hbar^2 f^{(3)} + \cdots$$

such that

$$f^{(1)}(1) = 0, \quad \Delta'' \circ \exp(f) = \exp(f) \circ \Delta', \quad \text{and} \tag{8.54}$$

$$\bigoplus_{n>k} S^n(U') \subset \mathrm{Ker}(f^{(k)}). \tag{8.55}$$

Notice that the first equation in (8.54) guarantees that such an f satisfies (8.53) so it belongs to $Lin^0_\hbar\big(S(U'), S(U'')\big)$ and thus the exponential in (8.54) exists. Another version of this definition was considered [13, §4.3] where (8.55) was replaced with a "dual" condition:

$$Im(f^{(k)}) \subset \bigoplus_{1 \leq n \leq k} S^n(U'').$$

[9]Convergence is always understood in the $\hbar$-adic topology.

As proved in [12, Corollary 33], *IBL*$_\infty$-algebras with the above morphisms form a category with the composition

$$f \diamond g := \log\big(\exp(f) \circ \exp(g)\big)$$

and the categorical unit

$$1_{S(U)} = \log\big(\mathbb{1}_{S(U)[[\hbar]]}\big).$$

As argued in [12], *IBL*$_\infty$-algebras form a subcategory of the still bigger category of *Markl–Voronov* algebras.

Example 8.5 Iterating the definition of the coproduct δ in $S(U)$, one obtains for the exponential in (8.54) the expression

$$\begin{aligned}&\exp(f)(u_1 \odot \cdots \odot u_n)\\&\quad= \sum \frac{1}{k!}\frac{\epsilon(\sigma)}{a_1!\cdots a_k!} f(u_{\sigma(1)} \odot \cdots \odot u_{\sigma(a_1)}) \cdots f(u_{\sigma(n-a_k+1)} \odot \cdots \odot u_{\sigma(n)}),\end{aligned}$$

where the summation runs over all permutations $\sigma \in \Sigma_n$, all $k \geq 1$, and all non-negative integers $a_1, \dots, a_k$ such that $a_1 + \cdots + a_k = n$. We recognize a formula in [3, Section 5].

Example 8.6 The category of *IBL*$_\infty$-algebras contains a non-full subcategory whose objects are L_∞-algebras as in Example 8.3 and morphisms are $\Bbbk$-linear maps

$$f : S(U') \to S(U'')$$

such that

$$f(1) = 0,\ \ \Delta'' \circ \exp(f) = \exp(f) \circ \Delta', \ \text{ and } \ Im(f) \subset U''.$$

Such a map automatically belongs to $Lin^0_\hbar\big(S(U'), S(U'')\big)$. We leave as an exercise to prove that

$$\exp(f) : S(U') \to S(U'')$$

is the unique extension of f into a morphism of unital algebras. We identify this result as the dual to that of [7, Remark 5.3] describing the category of L_∞-algebras and their (weak) L_∞-morphisms.

8.4 Comments and Remarks Related to Part I

In Sect. 6.1, we introduced cyclic operads. The simplest example is the cyclic commutative operad of Example 6.1. Another example is the cyclic associative operad. However, this one is described only later, cf. Sect. 6.2, Example 6.18, as a symmetrization of the non-Σ cyclic associative operad, Example 6.17. Further example is the cyclic endomorphism operad in Example 6.6 and its dual version, cf. Example 6.7. An important point is that each cyclic operad is a quotient of the free cyclic operad on some cyclic module, see Proposition 6.3. From the point of view adopted in Part I, cf. Sects. 3.8 and 4.3, cyclic operads are relevant to the classical string field theory, the commutative case to closed strings, the associative to open strings.

In Sect. 6.2, we introduced the non-Σ cyclic operads. The simplest example is the non-Σ cyclic associative operad in Example 6.17. As already mentioned, the cyclic associative operad is its symmetrization, cf. Example 6.18.

In Sect. 6.3 (cyclic) operad algebras were introduced. Such an algebra is a morphism from a cyclic operad to the endomorphism operad. This led us to non-unital Frobenius algebras in the associative case in Example 6.21 and to their commutative versions in the commutative one in Example 6.22. These are the familiar structures from the two-dimensional topological quantum field theory.

In Sect. 6.4 we discussed modular operads. Again, the simplest example is the modular commutative operad of Example 6.24. It is the modular completion (envelope) of the cyclic commutative operad. From the point of view of Part I, Sect. 3.8, it is the operad relevant to the quantum closed string field theory. This is the reason why we called it also quantum closed operad through the text. As for the modular associative operad (quantum open operad), its explicit description preceding Theorem 6.1 is rather complicated. Nevertheless, it is defined abstractly as the modular completion of the cyclic associative operad. Another possible description is as a symmetrization of the modular completion of the non-Σ cyclic associative operad. These two operads can be combined rather straightforwardly into the quantum open-closed operad. This is why we did not describe this operad explicitly.[10] Modular version of the endomorphism operad is used for defining representations. However, we don't get new examples of algebras from representations of modular operads. A cyclic operad and its modular envelope have the same categories of representations, due to the universal property of the modular completion.

To obtain homotopy algebras as representations of operads, and hence make a direct contact with string field theory and quantum field theory in general, we need the cobar construction and its modular analogue, the Feynman transform. Operads resulting from this constructions are, however, not modular operads any more. They are odd (twisted) modular operads. Their structure operations are of degree one, in

[10]Neither we discussed, which is much more delicate, its description in terms of a modular completion.

contrary to the modular operads having operations of degree zero. This unexpected degree is related to the degree of the BV bracket and BV operator Δ, which are inherently present in the construction of an algebra over the Feynman transform of a modular operad. Hence, in Sect. 6.5 we discussed odd modular operads. An example directly relevant to quantum field theory and string field theory is the odd modular version of the endomorphism operad. This is an operad based on a differential graded vector space equipped with a degree -1 symplectic form. In Part I, we have met examples of these spaces, namely the representation space of the first quantized point particle and of the closed, open, and open-closed CFT theory representation spaces, respectively. The differential was the BRST operator and the odd symplectic form was related to the product on the respective representation spaces, e.g. BPZ pairing for strings. Further examples we have met explicitly in Part I were the odd modular operads of chain complexes on the moduli spaces of Riemann surfaces associated with (open, closed, open-closed) string field theories. As also discussed in Part I, conformal field theory provides an example of an odd modular operad morphism going from the moduli space operad to the endomorphism operad.

In Sect. 7.2 we introduced the Feynman transform of modular cooperads which are objects dual to modular operads, in order to obtain an important class of odd modular operads. As for Part I, the most relevant modular cooperads come from dualizations of particular modular operads. The construction is an analogue of the classical cobar construction for cyclic cooperads. Although we mentioned the cobar construction at the beginning of the present section merely to motivate the Feynman transform, it is of its own interest and we will comment on it later. Nevertheless, we did not give a detailed description of the cobar construction. This can be recognized as the genus zero part of the more general case of the Feynman transform. The important point is Proposition 7.2 describing explicitly an algebra over the Feynman transform of a modular cooperad.

In Sect. 8.2, the relevance of operads in the description of string field theory became transparent. Here, the Feynman transform was applied to the case, when the modular cooperad is the component-wise linear dual of a modular operad.[11] In this instance, a morphism from the Feynman transform into an odd modular operad can be expressed directly using the structure of the original modular operad. It explicitly leads immediately not only to a BV structure, see Theorem 8.1, on the space of invariants (8.4), but also to a characterization of the corresponding morphism as a solution to the quantum BV master equation on this space, see Theorem 8.2. The BV bracket and Δ have their origins in the operadic operations of both the original modular operad and the (odd modular) endomorphism operad, cf. Proposition 8.1. The $S(n, g)$-part of the solution S to the master equation comes from evaluating the morphism on the (dual) of the (n, g) dg-vector space component of the original operad. We have met this through Part I. All decompositions of moduli spaces mentioned there are examples. More concretely, let us consider, for instance, closed

[11]In various places in Part I we would refer to this as to a Feynman transform of the modular operad itself. We will continue doing this also in the rest of this section.

strings. The geometric vertices of the closed string theory v_n^g define a solution to the master equation (3.17) on the space of (invariant under the permutations of punctures) chains on the moduli space of closed Riemann surfaces with punctures. They are the same thing as a particular morphism from the Feynman transform of the modular commutative operad to the odd modular operad of the chain complex on the moduli space. A two-colored version of that construction combining modular commutative and associative operads would give geometric vertices for the open-closed string field theory.

The BV formalism, as it is know from textbooks, can be recognized as an instance of Proposition 7.2 applied to a representation of the Feynman transform of the modular commutative operad in Sect. 8.2. In this case, we obtain a loop homotopy algebra as described in Theorem 8.3. In the physics language, fields are coordinate functionals on the dg-vector space (V, d). These would correspond to a full BV theory. The fields of degree 0 correspond to the original fields of the starting quantum field theory, fields being of positive degrees, anti-field of negative ones. Interaction vertices in the quantum BV action are the degree 0 totally (graded) symmetric functions f_n^g. The quantum BV master equation is recognized in (8.37). Here, d corresponds to the BRST operator. The role of degree one d-closed symmetric element $\sum s_i' \otimes s_i''$ in (8.37) is twofold. In the first term on the right-hand side of the equation, it gives the BV operator Δ. In the second one, it gives the BV bracket $\{-, -\}$. It is the same quantum BV master equation as the equation satisfied by the closed string field theory action in Part I, Sect. 3.4.

Obviously, a variant of Theorem 8.3 can be given in the case of modular associative operad and the quantum open-closed operad, the latter leading to the open-closed quantum BV master equation for action (5.1).

It should be now obvious, by looking at the $g = 0$ part of the above construction, that representations of the cobar construction for cyclic operads give the ordinary homotopy algebras. For example, in the case of the cyclic commutative operad we get cyclic L_∞-algebras whereas in the associative case we get cyclic A_∞-algebras. The corresponding algebraic structures are equivalent to solutions to the corresponding classical BV master equations. These are the algebraic structures discussed in relation to the point particle in Sect. 2.3 and in relation to classical string field theories, for instance, in Sects. 3.6 and 4.2, cf. also Sect. 5.2 for the classical open-closed case.

Finally, in Sect. 8.3, we reviewed IBL_∞-algebras, a common generalization of loop homotopy algebras and Lie bialgebras. They can also be understood as algebras over the cobar construction of properads. We however limited ourselves to their direct algebraic description without going into their properadic origins. They are generalizations of loop homotopy algebras in the following sense: in place of the second order, degree one, nilpotent differential operator $\hbar\Delta + \{S, -\}$, now we consider higher order, degree one, nilpotent differential operators. Correspondingly we have brackets with several inputs and outputs. In particular, we can have an ordinary bracket and cobracket as in a Lie bialgebra. The IBL_∞-algebras were relevant to our discussion of the open-closed string field theory in Sect. 5.2.

Although the IBL_∞-algebras used there are very particular ones, the one describing closed strings corresponds to a loop homotopy algebra and the one used to incorporate the open strings is an ordinary *IBL*-algebra, the full open-closed string field theory comprises their morphism in the IBL_∞ world. If we write the action as the sum $S_c + S_{oc} + S_o$, where S_c and S_o are the closed and open parts of the action, respectively, this morphism corresponds to the vertices in the S_{oc} part. Complementary to our description of IBL_∞ in Part II, we reviewed them also in Appendix A in a form directly used in our physics discussion of Part I.

References

1. Barannikov, S.: Modular operads and Batalin-Vilkovisky geometry. Int. Math. Res. Notices **2007**(19), Art. ID rnm075, 31 (2007). http://dx.doi.org/10.1093/imrn/rnm075
2. Bashkirov, D., Voronov, A.A.: The BV formalism for L_∞-algebras. ArXiv e-prints (2014)
3. Cieliebak, K., Latschev, J.: The Role of String Topology in Symplectic Field Theory. CRM Proc. Lecture Notes, vol. 49, pp. 113–146. American Mathematical Society, Providence (2009)
4. Cieliebak, K., Fukaya, K., Latschev, J.: Homological algebra related to surfaces with boundary. ArXiv e-prints (2015)
5. Kassel, C.: Quantum Groups, Graduate Texts in Mathematics, vol. 155. Springer, New York (1995). http://dx.doi.org/10.1007/978-1-4612-0783-2
6. Kravchenko, O.: Deformations of Batalin-Vilkovisky algebras. In: Poisson Geometry (Warsaw, 1998), Banach Center Publ., vol. 51, pp. 131–139. Polish Academy of Sciences, Warsaw (2000)
7. Lada, T., Markl, M.: Strongly homotopy Lie algebras. Commun. Algebra **23**(6), 2147–2161 (1995). http://dx.doi.org/10.1080/00927879508825335
8. Markl, M.: Models for operads. Commun. Algebra **24**(4), 1471–1500 (1996). http://dx.doi.org/10.1080/00927879608825647
9. Markl, M.: Loop homotopy algebras in closed string field theory. Commun. Math. Phys. **221**(2), 367–384 (2001). http://dx.doi.org/10.1007/PL00005575
10. Markl, M.: Operads and PROPs. In: Handbook of Algebra, vol. 5, pp. 87–140. Elsevier/North-Holland, Amsterdam (2008). http://dx.doi.org/10.1016/S1570-7954(07)05002-4
11. Markl, M.: On the origin of higher braces and higher-order derivations. J. Homotopy Relat. Struct. (2015). https://doi.org/10.1007/s40062-014-0079-2
12. Markl, M., Voronov, A.: The MV formalism for IBL_∞- and BV_∞-algebras. Lett. Math. Phys. **107**(8), 1515–1543 (2017)
13. Münster, K., Sachs, I.: Quantum open-closed homotopy algebra and string field theory. Commun. Math. Phys. **321**(3), 769–801 (2013). http://dx.doi.org/10.1007/s00220-012-1654-1

Index

© Springer Nature Switzerland AG 2020
M. Doubek et al., *Algebraic Structure of String Field Theory*, Lecture Notes in Physics 973, https://doi.org/10.1007/978-3-030-53056-3

Printed in the United States
By Bookmasters